D1260101

Yong-Qi Chen
Yuk-Cheung Lee (eds.)

Geographical Data Acquisition

SpringerWienNewYork

Prof. Dr. Yong-Qi Chen

Department of Land Information and Geo-Informatics, The Hong Kong Polytechnic University, Kowloon, Hong Kong

Prof. Dr. Yuk-Cheung Lee

Department of Geodesy and Geomatics Engineering, University of New Brunswick, Fredericton, NB, Canada

Printed in Austria

Typesetting: Scientific Publishing Services (P) Ltd., Madras
Printing: MANZ CROSSMEDIA, A-1051 Wien

Printed on acid-free and chlorine-free bleached paper
SPIN: 10690271

With 167 Figures

CIP data applied for

ISBN 3-211-83472-9 Springer-Verlag Wien New York

Preface

This is a book about techniques used in the acquisition of geographical data. The target audience is students and professionals using geographical information systems who want to go beyond the operation of the software and discover the general principles of how raw geographical data are acquired. By "raw" data we mean data acquired directly from the field, from photographs, or from maps but which has not been edited or structured for database storage. With this in mind, we have placed a heavier emphasis on geo-referencing and data acquisition techniques, making the coordinate reference framework an important link tying the chapters together. In writing this book, we have adopted a Scientific American-type style, which appeals to the technically curious layperson.

This is more than just a collection of articles, this is a textbook written jointly by several people. The coordination required for such an approach has made the production of this book much more difficult. The authors are predominantly faculty members of the Department of Land Surveying and Geo-Informatics at The Hong Kong Polytechnic University. We had hoped that this close proximity of authors could help us better coordinate the contents and ensure some consistency in style. This approach has certainly helped, particularly in allowing us to conduct the many meetings to discuss, review, and write the chapters together

The best way to write such a book is perhaps to let an expert pass on the technical knowledge to someone new to the field who then does the writing. Time constraint did not allow us the full freedom to follow this path, but we have adopted this approach in writing parts of the book that we considered particularly difficult to understand.

Dr. Zhilin Li was the one who had the vision for such a book in the first place. Other than contributing to the chapters, he has actively participated in the editorial and review process. Ms. Wendy Wells of the University of New Brunswick in Canada helped to polish the writing. Thanks are also due to all authors, whose patience, capacity to take criticism, and commitment to meet deadlines have singularly brought this rather difficult project to a completion within the allocated time.

The Editors

Contents

List of Contributors

Yong-Qi CHEN	Chair Professor and Head of the Department of Land Information and Geo-Informatics The Hong Kong Polytechnic University Kowloon, Hong Kong
Yuk-Cheung LEE	Professor Department of Geodesy and Geomatics Engineering University of New Brunswick Fredericton, NB, Canada
Jason C.H. CHAO	Assistant Professor Department of Land Information and Geo-Informatics The Hong Kong Polytechnic University Kowloon, Hong Kong
Xiao-Li DING	Associate Professor Department of Land Information and Geo-Informatics The Hong Kong Polytechnic University Kowloon, Hong Kong
Bruce KING	Assistant Professor Department of Land Information and Geo-Informatics The Hong Kong Polytechnic University Kowloon, Hong Kong
Kent LAM	Assistant Professor Department of Land Information and Geo-Informatics The Hong Kong Polytechnic University Kowloon, Hong Kong

Steve Y.W. LAM
Lecturer
Department of Land Information and Geo-Informatics
The Hong Kong Polytechnic University
Kowloon, Hong Kong

Zhilin LI
Associate Professor
Department of Land Information and Geo-Informatics
The Hong Kong Polytechnic University
Kowloon, Hong Kong

Esmond MOK
Associate Professor
Department of Land Information and Geo-Informatics
The Hong Kong Polytechnic University
Kowloon, Hong Kong

Lilian PUN
Assistant Professor
Department of Land Information and Geo-Informatics
The Hong Kong Polytechnic University
Kowloon, Hong Kong

Günther RETSCHER
Assistant Professor
Department of Engineering Geodesy
Vienna University of Technology
Gusshausstrasse 27–29 (E 127/2)
A-1040 Vienna, Austria

Qiming ZHOU
Associate Professor
Department of Geography
Hong Kong Baptist University
Hong Kong

1 Geographical Data and Its Acquisition

Yuk-Cheung Lee

1.1 Introduction

This chapter outlines the steps involved in the planning and acquisition of raw geographical data. It is both an introductory and a concluding chapter. As such, it might be useful to read this chapter first and then return to it after you have finished the others.

We consider raw data as those collected directly from the field or source documents, such as maps and aerial photographs, but unprocessed for error correction and cartographic enhancements. Acquisition of raw geographical data is the theme of this book. In this chapter, we will give an overview of the methods, provide a frame to tie the chapters together, and fill in some of the gaps that cannot be addressed appropriately in the individual chapters.

1.2 The Nature of Geographical Data

We collect data for a purpose. Geographical data acquisition is no exception, and the purpose it serves will ultimately determine the method to be used, the cost of the process, and the quality of the data acquired. Some authors prefer to regard *geographical* as a special case of *spatial* in that only the latter addresses the earth. We use the two terms interchangeably in this book most of the time, but we favour the use of the term spatial when only the geometrical properties are involved.

One of the main reasons for acquiring geographical data is to produce maps or, in modern terms, geographical databases. Analogue maps have served three functions: to store geographical data in graphic form, to display geographical data, and to support geographical analysis. Geographical databases in digital form serve similar purposes: to store geographical data and to support analysis. These databases, unlike maps, are not displays themselves.

This difference has affected the way data are organized. For instance, data on maps are cartographically symbolized to carry additional information we now call *attributes*. Digital data need not be symbolized for analytical purposes because attribute information can be carried as part of the database. If a visual image is required, digital data are then plotted on the screen or on paper. At that time, cartographic symbolization will be applied.

Conventional maps are results of interpretation, classification, geometric delineation, and cartographic enhancement. These kinds of data have an analogy in geographical databases and are called *vector data*. Although vector data have been interpreted, classified, and delineated, they need not be cartographically enhanced until display time. In conventional mapping, we also use image data, such as photographs, that have not gone through these processes. A combination of a line map and a photographic image, the *orthophotomap*, illustrates how the two types of data complement each other. In geographical databases image data are called *raster data*.

Geographical databases go beyond conventional maps and orthophotomaps to include data of higher dimensions and communication media other than graphics and images. The choice of data acquisition techniques is therefore more varied. Regardless of the nature of the geographical database to be created and the technological differences of the various methods, data acquisition consists of the following steps:

a) Define the nature and scope of the database.
b) Identify the types of features to be acquired.
c) Design the geographical database to contain the data.
d) Choose the method of data acquisition.
e) Acquire the data.

1.3 Define the Nature and Scope of the Database

A geographical database, like a map, is created for a group of applications with similar requirements. The first step in creating a database is to identify the need for it, the way it will be used, its size, the data source, and the amount of time available to carry out the project, and the budget. Many of these factors are related to each other. For instance, different sources of data produce data of different formats that in turn will affect the data volume and the cost of storage. The amount of time available and the budget will sometimes dictate the choice of data source.

In a top-down approach, this is design of the database at a very high level. The details of the steps involved will be discussed in sections 1.4 and 1.5.

1.4 Identify the Types of Features

For both maps and digital databases, geographical data acquisition is a process of abstraction, a process through which the highly complex world is generalized and simplified to a manageable level. In fact, this is true for the creation of any database, geographical or otherwise.

The world we live in is highly complex. An application requires data from only a part of it, sometimes called the *miniworld* [Elmasri and Navethe, 1994]. The first step in the abstraction process is to select those aspects of the real world that are usable by an application. During the selection, we consider the features, their characteristics, and their relationships with each other.

Features can be tangible "things", such as buildings, or intangible phenomena, such as the migration path of birds. We understand features here to mean geographical features (Figure 1.1), which *could* be given a location on Earth. The process of associating a feature with a location is called *geo-referencing* through a coordinate system. There are many coordinate systems used in daily life, some of which are not related to Earth at all. An example is the Red-Blue-Green (RBG) coordinate system for describing colour. For the description of geographical features, we are interested in only those coordinate systems related to Earth. This includes the geographical coordinate system of latitudes and longitudes (see Chapter 2) as well as plane coordinate systems based on map projections (see Chapter 4).

Features, being individual entities, are unique. Their properties are described by a set of *attributes*. For instance, the attributes of Hong Kong Island include its name, its surface area, the length of its coastline, its highest elevation, and so on. Some of the attributes have ties to coordinate systems and some have not. For example, the length of the coastline of Hong Kong Island could be derived from a list of points defining the coastline, and the points in turn are each defined by a set of coordinates. We will call these properties *geographical attributes*. A *non-geographical attribute*, such as the name of the island with the value "Hong Kong Island," is one that cannot be derived from coordinates. Note that we make a distinction between an

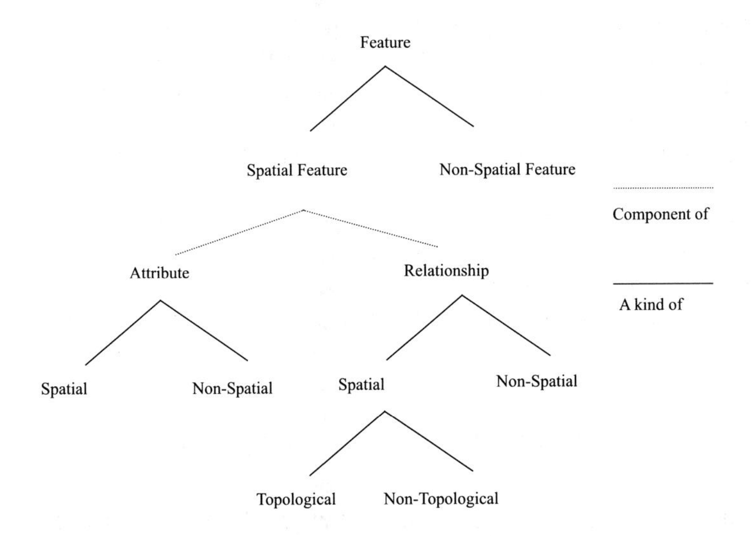

Fig. 1.1 Anatomy of spatial features

attribute (such as name) and its *attribute value* (such as "Hong Kong Island"). No two features should have the same attribute values because then they would not be unique.

Features never exist in isolation and are always related to each other. A *relation* describes how two features are related. Relations can be spatial or non-spatial in nature as well. *Spatial relations*, like spatial attributes, *could* be derived from coordinates. For example, the fact that two cities are 10 km apart (a relation of distance) could be derived given the location of the two cities. A *non-spatial relation*, such as Hong Kong is a Special Administrative Region of China, cannot be derived from coordinates alone.

Spatial relations are of two main types. The first type is *topological relation*, which is not affected by continuous transformation (distortion, deformation, etc.) of the coordinate system. A *continuous transformation* is like stretching features drawn on a rubber sheet without tearing and folding. The fact that Hong Kong Island is on the Pacific Ocean is topological. First of all, this relation is spatial because given the outline of Hong Kong Island and the outline of Pacific Ocean, one can derive that the island is within the ocean. This relation is topological because if we distort the map showing the island and the ocean, this enclosure relation will not change. The second type of spatial relation is distance relation, which is *non-topological* because it is affected by distortion.

In traditional mapping, we are rarely concerned with the capture of spatial relationships among features because they could be derived by the user when needed as long as the map is correct. We call these *implicit relations*. In a geographical database some of the spatial relations are stored explicitly. *Explicit relations*, therefore, are those that had been predetermined and stored as part of the data. The main reason for storing explicit relations is to speed up the process of a query because it is faster to retrieve stored data than to derive new information.

It is useful to identify features that share the same attributes (not attribute values) and assign a *feature type* to them. The relations between feature types are called *relation types*, while those between features are called relations. For example, a relation type called enclosure relates all lakes to their enclosing water bodies. A relation, also called enclosure, relates a particular lake to a particular enclosing water body.

The objective of this step is to identify the features types, their attributes, and the explicit relation types to be acquired for a particular database. This basically completes the abstraction process of generalising and simplifying the world for data acquisition.

1.5 Design the Geographical Database

An analogue map, being storage of geographical data, serves the purpose of a geographical database. Hence the notion of database design, which is the design of the "appearance" of data and the "container" for it, also applies. These design elements include surveying and mapping specifications such as accuracy of the data, scale, map projection, geometrical representation of features, symbology, rules of cartographic generalisation, and medium for the map or image.

The design of a geographical database is a more complex task than the design of traditional maps because of the number of parameters involved. Most of the design elements of conventional mapping apply, plus additional ones to handle the digital nature of data. For example, we need to design the coding standards for the attributes, the organization of the data in the database, and the input data format.

The details of geographical database design are beyond the scope of this book. We will consider in the chapters, however, the important characteristics of data that could affect geographical data acquisition, such as accuracy, coordinate system, and geometrical representation of features.

1.6 A Survey of Data Acquisition Methods

Ground-based data acquisition methods (Chapter 6) use direct observations to measure the position of objects. The surveying equipment is placed either directly at the point to be measured or within sight of it. Commonly used equipment includes total stations (Chapter 6) and Global Positioning System (GPS) receivers (Chapter 7). These methods are generally more accurate, more labour intensive, and more time consuming than air-based methods.

Between a total station and a GPS receiver, the former is usually more accurate. Total stations for engineering works typically will provide an accuracy of five seconds of arc in angular measurement (equivalent to an angle subtended by 2.4 cm at 1 km). For distance measurement, they are typically accurate to five parts per million, or 5 mm in 1 km. The accuracy of GPS receivers depends on a number of factors described in Chapter 7, but it can range from sub-centimetre to over one hundred metres. The major advantage of GPS surveying is that it is faster and less labour intensive than traditional high precision surveying using total stations. In addition, intervisibility between stations is not needed.

Air-based methods were devised to make position measurements easier to perform but at the expense of accuracy. These methods do not use direct observation, but produce an image of the area upon which to base the measurements. Photogrammetry is an early air-based method still very much in use for topographic mapping. It traditionally uses aerial photographs taken at altitudes of several kilometres, but larger scale aerial photographs taken at lower altitude are being used for engineering applications. Measurements are performed either on single photographs, which are geometrically distorted (Chapter 10), or on stereo models that are geometrically correct models of the real world (Chapter 11).

Remotely sensed imageries from satellites are basically aerial photographs taken from a much higher altitude (hundreds of kilometres) and with a spectrum much wider than the visual one. This allows us to interpret and identify geographical features and phenomenon not easily seen by the naked eye or photography (Chapter 12).

Since the image is a reduction of the real world, the measurement of features on it could be done much faster. On the negative side, an image cannot show Earth in full detail and some details might be obscured by others, thus rendering the air-based methods less reliable and less accurate. It is sometimes necessary to verify the nature and position of features observed on the image by a process called *ground truthing*.

The accuracy of data from these images depends on the height of the sensing platform, the quality of the imaging device, the stability of the imaging platform, and the quality of the measuring device. The height of the sensing platform affects the scale of the photography and the resolution of the image. Because of the flying height, the resolution of satellite imageries is usually lower than that of aerial photographs. For high quality aerial photographs on film, a resolution of 125 lines per millimetre can be reached. This is equivalent to about 0.3 m on the ground for a 1:40 000 aerial photograph commonly used for topographic mapping at 1:25 000. The resolution of satellite imagery is mostly in metres, and high-resolution imagery up to 1 m is now commercially available. It should be noted that airborne sensors carried by airplanes and even helicopters flying at lower altitudes can produce multi-spectrum imageries with very high resolution.

The quality of the imaging device affects the resolution of the resultant image and its geometry. Resolution determines the ability to identify with confidence the required feature on the image. High resolution, however, does not guarantee high accuracy. A high-resolution image can be geometrically unstable, producing distortions that change the relative position of features, thus affecting measurements such as position, length, and area. A poorly calibrated device, although of high resolution, can produce geometrically distorted images. Distortion can also be caused by the instability of the imaging platform, such as the airplane not flying along a horizontal path. These geometric distortions can be reduced to a certain extent given the proper method of geo-referencing, as explained in Chapters 3, 10, and 11.

The quality of the measuring device affects position measurement on an image. An analogue photogrammetric plotter, for example, can yield an accuracy of about fifteen microns at the image scale, while a more advanced analytical plotter could yield an accuracy of about three microns.

A rather unique source of data is maps. They are similar to images in that they are reduced models of the real world. They are different from images because they contain interpreted data, thus making ground truthing unnecessary. Because they are secondary data normally derived from images in the first place, they are even lower in accuracy. In a high quality topographic map, some 90% of its well-defined features are within 0.5 mm of their true planimetric position at the map scale. That translates to 12.5 m on the ground on a 1:25 000 map. Vertical accuracy is about half of the contour interval. The basic techniques of capturing data from hardcopy maps are manual digitising and automatic scanning. The much slower manual method involves an operator tracing lines on a map at a typical speed of 1.5 mm per second and producing vector data. Automatic scanning produces raster data that must be processed if they are needed to support analysis. The complete automation of processing raster data for analysis is extremely difficult (Chapter 5).

Data capture on both maps and images from air-based methods is much faster than ground-based methods. A typical tracking speed of 1.5 mm per second on a 1:25 000 map is equivalent to a ground speed of 135 km per hour unaffected by traffic and the type of terrain.

A hydrographic survey also uses "remote sensing" techniques to obtain measurements of the seabed terrain using equipment such as echo sounders (Chapter 8). According to the International Hydrographic Organization's (IHO) specifications, the better hydrographic charts for shallow water areas provide a horizontal accuracy of about two metres and a depth accuracy of about 0.25 m (Chapter 8). A hydrographic survey typically generates large number of soundings, which are depth measurements.

1.7 Geo-Reference Data

In this step, we will identify features, measure their locations, collect data for their attributes, record their explicit relations, and eventually import these data into the database. Specific techniques for different data capture devices have been described in various chapters of this book.

An important step in data acquisition is the geo-referencing of data, a process that relates raw data to a useful coordinate system. Other than the geographical (geodetic) coordinate system giving latitude and longitude of a point in degrees, we use plane coordinate systems based on map projections (Chapter 4). Unlike the geographical coordinate system that is based on an ellipsoidal earth, these plane coordinate systems are based on a flat surface although they are actually the transformation of geographical coordinates onto a flat plane. Plane coordinates are often in metres or other units of length measurement commonly used on the ground.

The geo-referencing process depends on the acquisition methods. When we perform a ground-based survey using satellite positioning techniques, the position of the surveyed points would have been geo-referenced by the associated software, and the reading of the coordinates is in geographical latitude/longitude or metres on a plane coordinate system.

When the ground-based survey was done using traditional instruments, such as transits and levels, some transformation of coordinate systems would be required. The polar coordinates resulting from the measurement of angles and distances would have to be converted to plane coordinates to fit those shown on maps. An important point to note when combining coordinates obtained using these methods and those from satellite positioning systems is the difference in height measurement. Satellite positioning systems use a different reference (called datum) than traditional surveying methods to determine height (Chapter 2), and the conversion between the references is not trivial. The same applies to the reference used in the two techniques to measure horizontal coordinates, except that the conversion in this case is well defined mathematically.

A single aerial photograph or satellite image contains distortions caused by terrain differences (Chapter 10). We can geo-reference it approximately using a simple procedure called rectification or registration. Using more complex calculations incorporating knowledge about the terrain, we can turn a single image into a geo-referenced orthoimage that behaves geometrically exactly like a map.

If we have a stereo-pair of images, we can form a stereo-model of the terrain that is free from terrain distortions. From that, a procedure called absolute orientation can be used to geo-reference the stereo-model to a plane coordinate system (Chapter 11). This is a standard procedure in setting up a stereo-model for photogrammetric operations.

To acquire data from a map, a digitising table is usually employed to trace data on a line map for conversion into a form compatible with the database. The coordinate system of the digitising table is unrelated to a map projection until we perform a registration (Chapter 5). After registration, all points measured on the digitising table will be converted automatically to fit the map system. If a map is scanned instead of digitised by hand, the end results are geometrically equivalent. Hence the technique of registration can also be used to geo-reference a scanned map.

1.8 Trends in Spatial Data Acquisition

Data acquisition has always been a bottleneck in the implementation of geographical databases. There are two aspects to geographical data acquisition: geometric and thematic. The challenge involved in eliminating or just reducing the data capture bottleneck lies in the attempts to automate the capture of both geometric and thematic data.

The geometric and thematic aspects are very much related. In delineating the geometric outline of a feature, we must first identify the boundary between two features (or themes). Alternatively, after identifying the feature to which every single location in a project area belongs, we can derive the outline of the features. In other words, geometric and thematic information share a dual relationship with each other.

Regardless of this dual relationship between the two aspects, we often classify data acquisition methods as either producing geometric or thematic information. Those techniques we normally consider as producing geometric information, such as photogrammetry, are predominantly manual and provide vector data. The automated techniques, such as feature classification as described in Chapter 12, are more effective in producing thematic information and use raster data exclusively. The basic principle of these classification techniques is to identify the special spectral characteristics of objects we are interested in and to classify each pixel according to its spectral value. The delineating of feature boundaries from raster data, as explained in Chapter 5, cannot yet be automated completely.

Automated feature extraction from aerial photographs and satellite imageries is actually trying to solve the two problems of delineating an outline and classifying a feature at the same time. Understandably, it is a very difficult problem. Even when human interpretation is involved, the identification of features on aerial photographs cannot be completely reliable as features are obscured by shadows, hidden by other features, or in some cases plainly deceptive. Automated feature recognition by computers is a task many degrees of magnitude more difficult. Research in this area has resulted in methods that use a multitude of parameters such as spectral value, texture, shadow, elevation, and terrain to help identify and classify features [Gurney, 1981; Harris and Ventura, 1995; Zhang, 1999]. A very important feature is the terrain surface itself. The automatic generation of a digital elevation model (DEM), which is a collection of points with known horizontal and vertical coordinates, to represent the terrain surface has been a major area of research and development.

A lot of attention in recent years has been given to the detection of vegetation and buildings, the two major features that hide the ground from the generation of digital DEM. Without their removal, the DEM generated will depict the tops of trees and the roofs of buildings. Vegetation has an "open surface" in that the ground can sometimes be seen in a vertical image through small openings, but buildings have "closed" surfaces. Vegetation has a rather fuzzy outline, while building corners are very well defined, and these two features have rather distinctive textures. Recent developments in Airborne Laser Scanners (ALS) [Ackermann, 1999; Baltsavias, 1999] have produced promising results in the removal of vegetation and buildings from airborne imageries to create a terrain surface. ALS can penetrate open surfaces to some extent and can automatically generate X,Y,Z coordinates to make the detection of buildings more reliable [Haala and Brenner, 1999].

The need for an increasing number of parameters calls for the integration of GIS, photogrammetry, and image processing technology to provide an environment that can facilitate geographical data acquisition. In an integrated environment, the GIS can provide data of known thematic values to help classify pixels of unknown thematic values.

The integration of technology goes beyond this to include data acquisition hardware as well, which is exemplified by the integration of GIS and the Global Positioning System (GPS). As explained in Chapter 7, GPS uses satellite positioning technology to pinpoint locations on Earth using portable receivers. This integration has given mobility to GIS software that effectively brings GIS to the field. Consequently, new breeds of data acquisition systems have evolved. One of them is the Mobile Mapping System (MMS) that uses multi-sensors and a satellite positioning system to automatically survey objects visible from a moving vehicle. These systems can offer sub-metre accuracy when installed on a vehicle moving at the moderate speed of 60 km/h. A portable version of it with less automation and lower accuracy is used extensively in field mapping involving small teams of not more than two persons. Supporting hardware used in this case, other than the essential GPS receiver, includes laser ranging devices for distance measurements accurate to about 1/1000 of distance, digital still cameras, and video recorders. Air-borne systems of similar function have been used increasingly to capture high-resolution imageries incorporated with positioning information.

Echoing the rapid development of orthophotomaps in the late 1970s because of the need for fast coverage of unmapped terrains, in recent years we have seen a surging popularity of digital orthophotography [Spradley, 1996; Baltsavias, 1996]. Back in the 1970s, orthophotographs could be produced automatically, and the softcopy equivalent of the early system is extensively used today. With the higher resolution of satellite imageries, we can expect the increasing use of orthophotographs for "rapid mapping." The automated creation of orthophotographs requires accurate DEM.

It appears that images will play a significant role in future geographical databases. Other than high altitude images, such as aerial photographs, we start to find many applications for ground images as well. Terrestrial photogrammetry and close-range photogrammetry [Clarke, 1995], the techniques of conducting surveys using land-based images, is the standard technique for many aspects of engineering surveying, urban surveying, and even medical applications. Together with digital imaging, these techniques find new applications in mobile mapping systems where image capture is done at the ground level. Images are also useful for capturing details that are too expensive for traditional surveying techniques to be used and for generating virtual reality presentations in an inexpensive way [Chapman and Deacon, 1998].

The growing need for images and other kinds of data, the multi-media explosion, together with the demand for rapid data capture will shape the development of geographical data acquisition in the future.

2 Coordinate Systems and Datum

Esmond Mok and Jason Chao

Geo-referencing is the technique of assigning location codes to points on Earth. There are many geo-referencing methods, including the use of landmarks and civic addresses. An effective system would provide a systematic way of yielding a unique code for each possible location. In other words, the location code of any position could be determined uniquely using a well-defined procedure. Moreover, given a code, the location it represents can also be derived. Both landmarks and civic addresses are not every effective geo-references because they are not unique.

When we talk about a space, we often talk about its dimension, which is the number of parameters needed to uniquely identify locations in the space. In a one-dimensional space analogous to a line, one parameter is enough. If we use one end of the line as the origin, then any position along the line could be represented by its distance from the origin. Such a parameter is called a *coordinate*. The measurement in distance is given a unit such as centimetres or kilometres. In a two-dimensional space, two coordinates are needed to obtain unique position codes. An extension of the one-dimensional method can be used here, giving rise to two distance measurements along two coordinate axes. In a three-dimensional space, three coordinates are required.

This chapter explains the characteristics of coordinate systems, the different kinds of coordinate systems, and their use in geographical data acquisition.

2.1 Coordinate Systems

A *coordinate system* has an origin, a set of independent axes, an orientation of the axes, a unit of measurement along the axes, and a method of reading coordinates. In short, it is a set of rules on how to assign coordinates to locations. Once the origin and orientation of the axes have been specified, we have a *reference coordinate system*. *Geo-referencing* is a technique of relating points on Earth to an Earth-based coordinate system. We have excluded here the discussion of coordinate systems that are not related to Earth, such as the RGB (Red-Blue-Green) coordinate system for colours.

2.1.1 Cartesian Coordinate System

A *two-dimensional Cartesian coordinate system* is defined by two mutually perpendicular axes. Mathematicians usually call them the X and Y axes. However, surveyors usually relate geographical features to the north direction, therefore in surveying and mapping applications the two axes point to the north and east, respectively. The intersection of the two axes is the origin of the coordinate system. This origin usually has zero value for both X and Y (north and east) directions. Figure 2.1 shows how point P is defined by projecting it onto the north and the east axes. If the distance from the origin to P′ is E_P, and that from the origin to P″ is N_P, then P is located at (N_P, E_P) or (E_P, N_P) depending on the convention used.

Following the above principle, three mutually perpendicular axes define a three-dimensional coordinate system. As shown in Figure 2.2, a point P is defined by the distance from the origin O along the X, Y, and Z axes. The horizontal position is defined by (X, Y) or (N, E) as in the two-dimensional case, and the height component yields the Z coordinate.

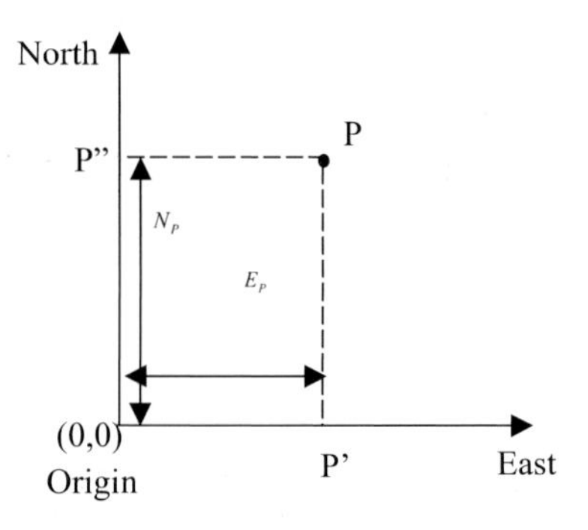

Fig. 2.1 Position of P defined by a Cartesian coordinate system

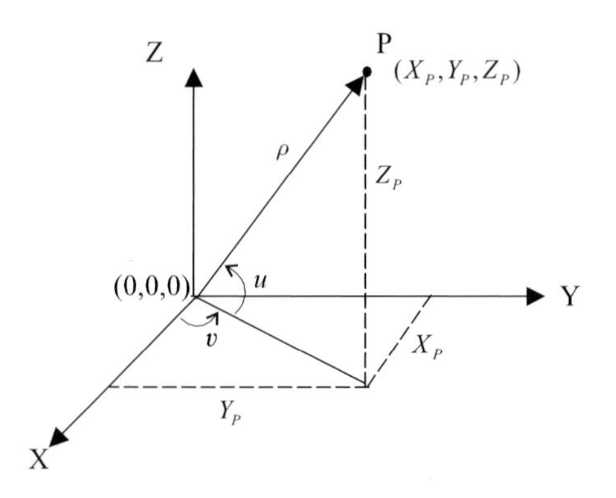

Fig. 2.2 A three-dimensional Cartesian coordinate system

The Cartesian coordinate system is very popular because the method of measuring the X, Y, and Z coordinates is very simple. They are no more than linear measurements along the three respective axes.

2.1.2 Polar Coordinate System

A *polar coordinate system* defines the location of a point by a combination of angle and distance. In the two-dimensional case, an angle θ and a distance ρ are used (Figure 2.3).

Similarly, for the three-dimensional case, the location of point P can be expressed in terms of angles (u, v) and distance ρ as shown in Figure 2.2. In this case, v is the angle measured anti-clockwise from the X-axis to the line containing the origin and the projection of P on the X-Y (horizontal) plane; u is the vertical angle measured from the X-Y plane to P; and ρ is the slope distance from O to P.

Polar coordinates are particularly useful for ground surveying using a theodolite and a distance-measuring device (Chapter 6). The objective here is to derive the position of P through observations at a known point O which may not be the origin (Figure 2.3). A theodolite will be placed at O to make the angular measurement θ and a distance-measuring device will be used to measure distance ρ. If we know the Cartesian coordinates of O, we can derive the Cartesian coordinates of P from the polar measurements (see section 2.1.3).

2.1.3 Conversion between Cartesian and Polar Coordinate Systems

Very often, it is necessary to convert positions represented by the Cartesian coordinates to the corresponding polar form, and vise versa. A practical example in ground surveying is given in the previous section.

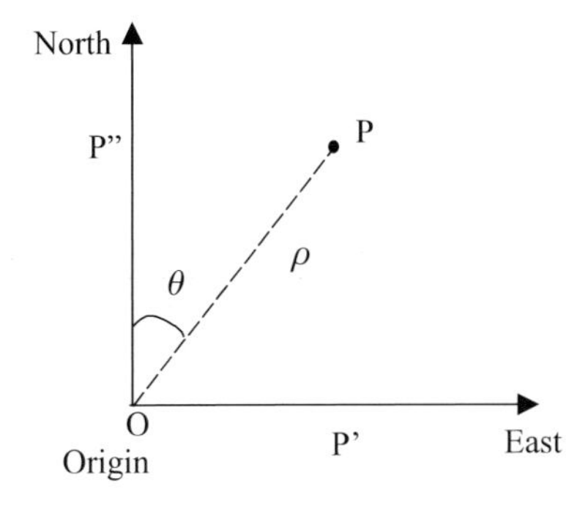

Fig. 2.3 Position of P defined in a polar form

The following formulae and calculation procedures convert between the polar and Cartesian forms.

$$
\begin{aligned}
X &= \rho \cos u \cos v, \\
Y &= \rho \cos u \sin v, \\
Z &= \rho \sin u \\
v &= \tan^{-1}\left(\frac{Y}{X}\right), \\
u &= \tan^{-1}\left(\frac{Z}{X} \cdot \cos v\right), \\
\rho &= \frac{Z}{\sin u}.
\end{aligned}
$$

In the above calculation, we assume that the polar measurements are from the origin. If the measurements are from a point with known (X_o, Y_o, Z_o) coordinates, then the true coordinate of the surveyed point is $(X_o + X, Y_o + Y, Z_o + Z)$.

2.2 Geoid and Mean Sea Level

A coordinate system is often applied to a space within which measurements are to be made. A three-dimensional system is applied to a three-dimensional space, and a two-dimensional system is applied to a two-dimensional space. For example, before we perform position measurements on a sheet of paper, we will first construct a two-dimensional coordinate system on it. If the surface used were highly irregular, the construction of a coordinate system over it and the subsequent reading of coordinates on it would be rather difficult, if not impossible. This is the problem with which we are faced when reading positions over the irregular surface of Earth.

If Earth were entirely covered by water, we would have a completely smooth surface with which to work. We assume here that external forces, such as tide, current, and wind, do not disturb this surface. Ground surveying is often performed on a horizontal surface defined by the plumb line of the equipment (Chapter 6). On a smooth surface completely covered by water, the plumb line will be everywhere perpendicular to the water's surface, which is Earth's surface in this case. If we further imagine that Earth is a sphere (which is about right for small scale operations) then a mathematical expression can be used to define its shape.

As we know, the shape of Earth is not that regular. Although 70% of Earth's surface is covered by water, the rest is covered by land mass of different densities. Furthermore, Earth's water surface is disturbed by external forces. The distribution of land masses affects the magnitude and direction of gravity, hence disturbing the direction of the plumb line on surveying instruments.

Let us explain this in more detail. Consider a point P with unit mass is located near Earth's surface. According to Newton's law of gravitation, point P would be attracted by the mass of Earth. This force vector shown in Figure 2.4 is denoted by $\vec{F}$. In addition, Earth completes one revolution in about 24 hours. This rotation generates the centrifugal force $\vec{C}$, pulling P outward. The resultant of these two forces is the gravity force $\vec{G}$; its direction is called the "plumb line" of P. If point P is at another position, then the non-uniform distribution of Earth's density will cause the change in magnitude and direction of $\vec{G}$.

The magnitude of $\vec{G}$ is the *potential energy* of P, or simply called "*potential*". Since P can be anywhere near or on Earth's surface, there is a "field of gravity" with different variations in the potential surrounding Earth. A surface formed by locations with the same potential is called an "equi-potential surface", or surface of equal gravity pull. Since the magnitude of gravity pull is a function of distance from the mass, an equi-potential surface can be regarded as a surface of constant "height" in a gravitational field. There are an infinite number of equi-potential surfaces covering Earth (see Figure 2.5). One with the "zero height" would be a convenient choice for measuring elevation. In the ideal Earth completely covered by water, the water's surface would be an ideal choice for zero height. This surface is called the *geoid.*

In a less ideal Earth, where the presence of landmasses of different densities disturbs the gravity field, an equi-potential surface around Earth

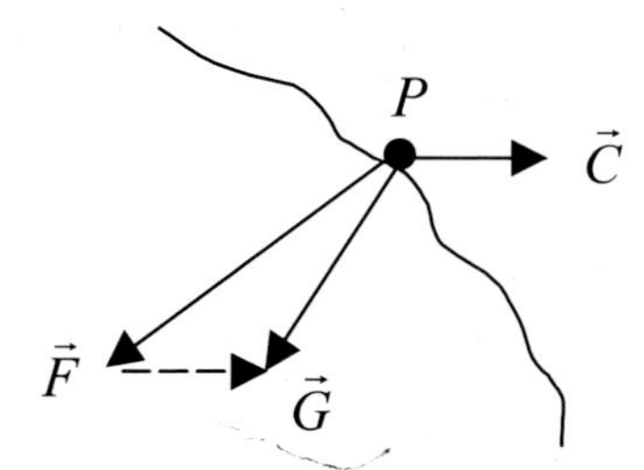

Fig. 2.4 Gravity is the resultant of the centrifugal and the gravitational forces

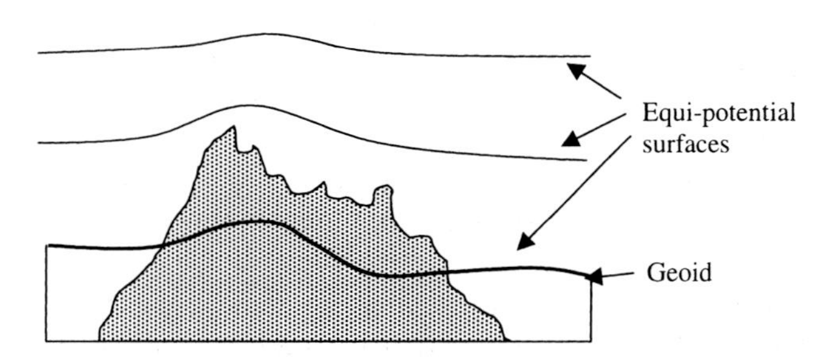

Fig. 2.5 Geoid surface is one of the infinite numbers of equi-potential surfaces

will be undulatory. Consequently, the geoid surface in reality is irregular in shape. Over oceans, however, this equi-potential surface will be very close to the water's surface. That gives us a convenient reference for determining the location of the geoid surface at coastal regions, and this is done by averaging sea level changes to yield the *mean sea level* after eliminating the effect of external forces on the water. It is normally determined by averaging many years of observation on the water level at a selected site, whereas recording of sea level data is achieved with an automatic recording tide gauge. The number of years of observation varies from place to place. For example, the United Kingdom's mean sea level was calculated from 6 years of hourly tide gauge readings recorded at Newlyn, Cornwall, whereas Hong Kong's mean sea level was determined based on 19 years of tide gauge records. Averaging of many years of sea level data can effectively reduce the periodic tidal effects as well as other random effects due to currents and wind (Chapter 8.2).

The mean sea level in reality does not extend onto land. It is, however, possible to mathematically approximate a mean sea level over the entire Earth's surface, yielding a continuous geoid surface. Because the change in gravity is slow from place to place, the geoid surface is smooth although irregular, making it quite difficult to be defined exactly using mathematical expressions. To visualize this surface, you can image that Earth's land mass is a porous sponge with the proper gravitational pull but which allows water to seep through and settle without friction. Alternatively, one can start with the smooth geoid surface on an ideal Earth completely covered by water, and by adding gravitational fields to simulate the pull of the land masses, an undulating geoid surface will be formed. Determining the actual shape of the geoid is a tedious and complex task, and it involves a large number of gravity observations. This is nevertheless an important task because the geoid, although rather abstract, can be determined, and its shape affects the operation of traditional surveying instruments.

The need for the geoid and its association with the mean sea level has to do with how surveying on the ground is being done traditionally. As will be explained in Chapter 6, ground-surveying instruments operate on the vertical and horizontal planes. The two planes are used to measure vertical angles, horizontal angles, and height differences. These instruments find the vertical and horizontal planes through the help of plumb lines and spirit

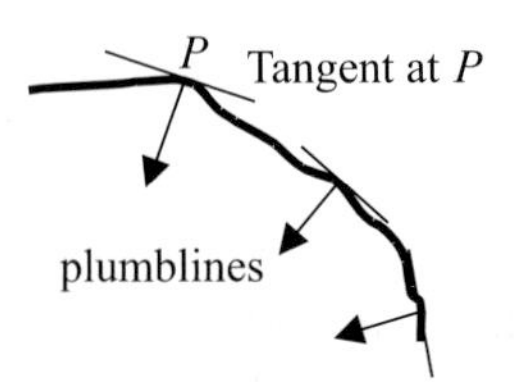

Fig. 2.6 Everywhere on the geoid surface is perpendicular to the direction of local gravity

levels. Both of them are sensitive to gravity pulls, and therefore traditional surveying instruments are actually built for the geoid. The plumb line, for example, is vertical to the geoid surface at that spot (Figure 2.6). The spirit level, on the other hand, reflects the direction of the equi-potential surface at that point. So all these measurements are being performed on the geoid surface that we can sense but cannot see. With accurate knowledge about the geoid, we can make these measurements more reliable.

The geoid is particularly important for measuring heights using traditional surveying instruments. The mean sea level is a very convenient origin for height measurements because it is a rather tangible mark for the beginning of dry land. Hence the mean sea level approximating the geoid would give us a reference for measuring height. Heights above the geoid are called *orthometric heights*, and they are those measured by levelling instruments described in Chapter 6. Another mathematical body called the ellipsoid (section 2.4) is also used for referencing elevations, e.g., in the case of satellite positioning systems (Chapter 7).

2.3 Datum

In theory, a Cartesian coordinate system can be used to describe positions on Earth's surface. To do this, we simply fix a Cartesian coordinate system in space, such as using the centre of a spherical Earth as its origin and orienting the axes accordingly. All coordinate measures are hence from the centre of Earth.

Such a system is not very useful for surveying and mapping. First of all, horizontal positions are confined to the surface of Earth, but a Cartesian system can generate points above and beyond this surface. Consequently, most of the points on a Cartesian system are beyond the range of mapping applications. Second, coordinates measured from the centre of Earth are of little practical meaning. Even for elevations, which are measured along a vertical line through the centre, we would still prefer height measurements relative to the surface of Earth and not its centre.

For these reasons, the concept of a datum was introduced. For vertical measurements, the mean sea level (the geoid) has been a popular *vertical datum* from which elevations and depths are referenced. For horizontal measurements, we fix a mathematical body of Earth in space using a Cartesian coordinate system. After that, a separate coordinate system is created over the surface of this body to generate horizontal coordinates. A mathematical earth body fixed in space makes up the *horizontal datum*.

A sphere is a first approximation to Earth (in fact, to the geoid), but for more accurate measurements, an ellipsoid is used (section 2.4). The horizontal coordinate system over the ellipsoid is called the geodetic coordinate system (section 2.4.1), a form of which is also known as the geographical coordinate system. The ellipsoid surface can also be used to define a vertical datum. It is important to note that datums are not unique so that many different ellipsoids can be used to approximate Earth (section 2.4).

Datums tend to evolve, as we know more about the true shape of Earth. For example, the North American Datum of 1927 (NAD27) was modified in 1983 to become NAD83 [Junkins and Garrard, 1998]. As a result of such a datum change, coordinates of points on Earth will change.

2.4 Ellipsoid

Although geoid is a more accurate representation of Earth's shape and a good datum for vertical measurements, it does not provide a suitable surface for horizontal position computations. This is because the geoid is irregular due to the non-uniform distribution of Earth's landmasses. It will be extremely difficult, if not impossible, to apply mathematical rules to calculate spatial positions of features on this surface. Therefore, a body that can be defined by mathematics and is very close to the shape of the geoid is used instead. Once this mathematical model is defined, positions of geographic features can be uniquely and clearly defined and geometric calculations can be performed.

The exact shape of Earth's surface is too complex to be represented by simple geometrical figures. For a rough approximation to the geoid, we can regard it as a sphere. The undulations of the topographic variations can be ignored for mapping horizontal positions because these variations are very small compared to Earth's size. If we shrink Earth to a ball of two metres in diameter, the topographic variation shown to scale would be no larger than 2.5 millimetres. For maps with a scale of 1:1 000 000 or smaller, the error caused by the spherical assumption is not noticeable, and a sphere with a radius of 6 371 km is a good approximation to Earth. This approximation is simple but it can represent only a very crude shape of Earth, therefore it is normally used for cartographic work where the accuracy requirement is low.

For a better approximation, Earth can be regarded as an *ellipsoid* obtained by rotating an ellipse along its short (semi-minor) axis. Figure 2.7 shows an ellipse, with the X and Z coordinate axes passing through the semi-major (the longer axis) and semi-minor axis. The shape and size of the ellipse are defined by the length of these two axes (a and b). With the knowledge of these two parameters, the *flattening* (f) and *eccentricities* (e and e') can be deduced. It should be noted that the ratio $1/f$ is usually adopted to represent flattening.

An ellipsoid is indeed a geoid of a smooth Earth if we remove the lateral variations in density due to landmasses and assume that density varies uniformly from the centre of Earth to its surface.

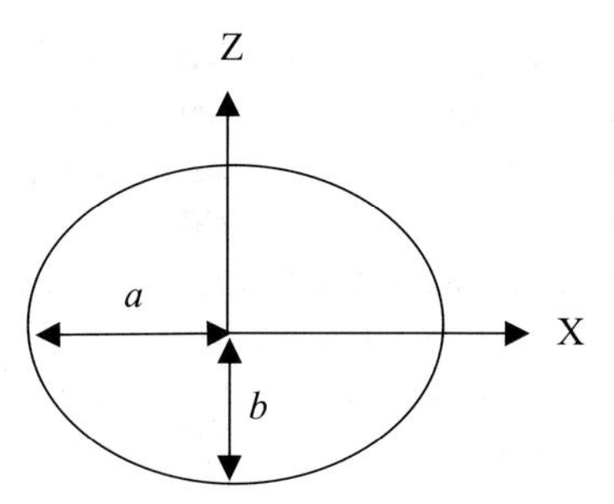

Fig. 2.7 The size and shape of an ellipse is determined by a and b

Some useful mathematical relationships of the basic elements a, b, f, e and e' are

$$f = \frac{a-b}{a}$$

$$e = \frac{\sqrt{a^2-b^2}}{a} = e'\sqrt{1-e^2}$$

$$e' = \frac{\sqrt{a^2-b^2}}{b}$$

$$e^2 = 2f - f^2$$

$$1 - e^2 = (1-f)^2$$

where
a is the semi-major axis
b is the semi-minor axis
f is the flattening
e is the first eccentricity
e' is the second eccentricity.

Flattening and eccentricities are also indicators of the shape of the ellipse. If a equals b, the ellipse shown in Figure 2.7 will become a circle. The flattening, as well as the eccentricities, will become zero. Actually, the difference between a and b for Earth is about 22 kilometres.

If the shape and size of an ellipse are assigned with appropriate values, and the ellipse is rotated about the Z axis, an oblate ellipsoid that closely fits Earth's surface can be defined (see Figure 2.8).

Being an approximation, different ellipsoids are used to map different parts of Earth. This is because Earth is not perfectly ellipsoidal and different ellipsoids will fit different parts of Earth better. For instance, the ellipsoid that has been used for North America for quite a long time was developed by Alexander Ross Clarke in 1866. The following are some of the common ellipsoids.

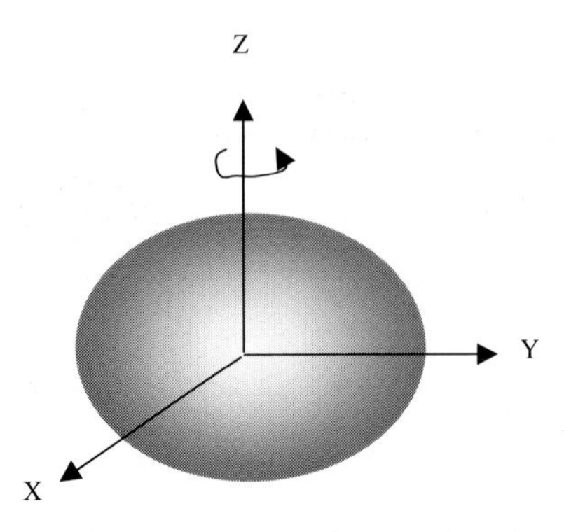

Fig. 2.8 An oblate ellipsoid is generated by rotating the ellipse in one cycle about the Z axis

Name	Date	Semi-Major Axis (m)	Semi-Minor Axis (m)	Usage
WGS84	1984	6378137.0	6356752.3	Global
GRS	1980	6378137.0	6356752.3	Global
Clarke	1880	6378249.1	6356514.9	Most of Africa, France
Clarke	1866	6378206.4	6356583.8	North America, Philippines
Hayford	1910	6378388.0	6356911.9	Hong Kong

It should be noted that the ellipsoid itself cannot provide a system to define positions because it can be anywhere and of any orientations in the three-dimensional space. An ellipsoidal coordinate system can be established by attaching an ellipsoid to the Cartesian coordinate system. For example, an ellipsoid attached to a Cartesian coordinate framework, would have the origin fixed at the centre of mass of Earth, the X-axis pointing to the Greenwich meridian, and the Z-axis pointing to the north pole of Earth. Given a sphere or an ellipsoid, with the specified coordinate system, we can then define a network of latitudes and longitudes (graticule) over it, thereby providing a framework to give each location on Earth a pair of coordinates. This is commonly known as the geodetic (ellipsoidal) coordinate system. Such a body that is used to define a reference system for coordinates is called a reference body, and such an ellipsoid is called a reference ellipsoid.

2.4.1 Geodetic (Ellipsoidal) Coordinates

An ellipsoidal coordinate system is either global or local, depending on whether the ellipsoid is used to best fit the whole world or a particular region of it. The local-fit ellipsoid aims at minimizing the separation between the

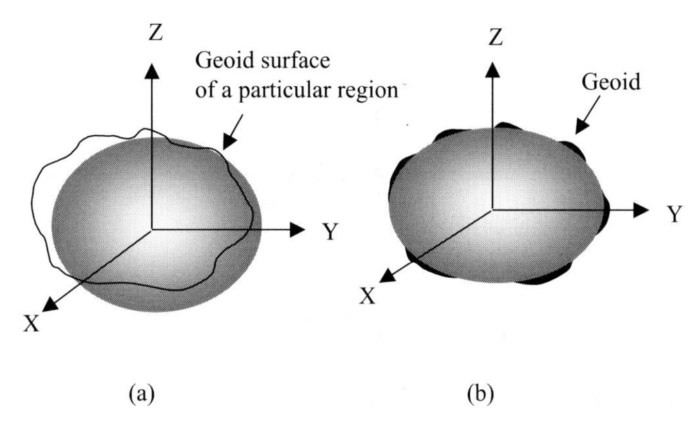

Fig. 2.9 Concept of local-fit and global-fit ellipsoids

geoid and the ellipsoid surface (usually called *geoid-ellipsoid separation*) of a particular region (see Figure 2.9(a)), while the global-fit ellipsoid is to make this separation the minimum over the whole world (see Figure 2.9(b)).

A global-fit ellipsoid usually attached to a Cartesian coordinate frame, would have the origin fixed at the centre of mass of Earth (i.e. *geocentric*), the X-axis pointing to the Greenwich meridian, and the Z-axis pointing to the North Pole. However, a local-fit ellipsoid may have the origin of the Cartesian coordinate frame offset from Earth's centre, and the three axes slightly deviate in angle from those used to fix a global-fit ellipsoid (Figure 2.10). Consequently, the same point P could have different coordinates depending on the type of ellipsoid used. It is possible to convert the coordinates from one system to another, however, as long as we know the parameters defining both systems.

The position of a point under an ellipsoidal coordinate system can be defined either in polar form as latitude, longitude, and height $(\varphi_P, \lambda_P, h)$, or in Cartesian coordinates (see Figure 2.11). The latitudes and longitudes are popularly known as the *geographical coordinates*.

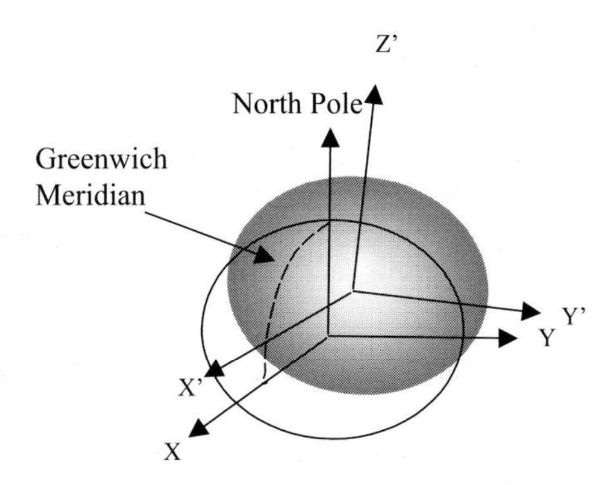

Fig. 2.10 The global (with axes X-Y-Z) and local fit (with axes X′-Y′-Z′) ellipsoids may have different orientation and origin definition

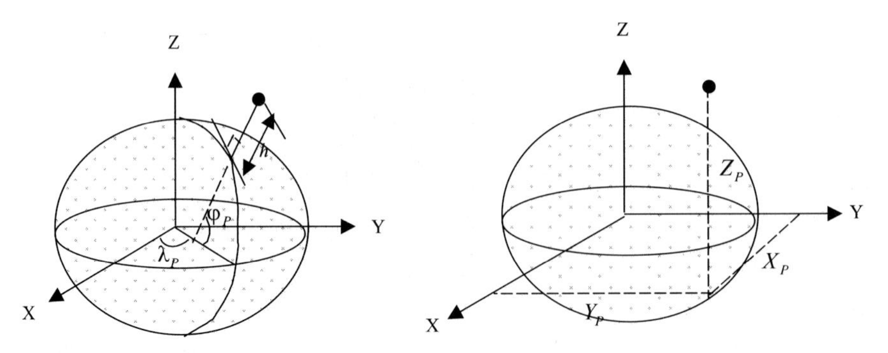

Fig. 2.11 Polar and Cartesian representation of a point in an ellipsoidal coordinate system

These two expressions can be converted from one to the other. That means, if the Cartesian coordinates of a position are known, the corresponding latitude, longitude, and ellipsoidal height can be determined, and vise versa. This system is a three-dimensional geodetic coordinate system, in which both horizontal and height positions are referred to the same ellipsoid. Some countries define horizontal and height using two separate reference systems with the mean sea level used for height. With the different reference systems for horizontal position and height, it is not possible to transform the features from latitude, longitude, and height to the three-dimensional Cartesian coordinates, unless the mean sea level height (the orthometric height) is first transformed to the *ellipsoidal height*, or height about the ellipsoid. Such transformation requires knowledge of the geoid.

Section 2.4.2 below discusses the conversion between geodetic and Cartesian coordinates. The element N in the conversion formula is the radius of curvature of a point in the prime vertical direction of an ellipsoid. Unlike a sphere the radius of which is a constant, the radius of a point on an ellipsoid will vary with position and direction. Figure 2.12 shows a portion of an ellipsoidal surface. Now let us consider that P and P′ are extremely close to each other. When PP′ are in the north-south direction, the radius of

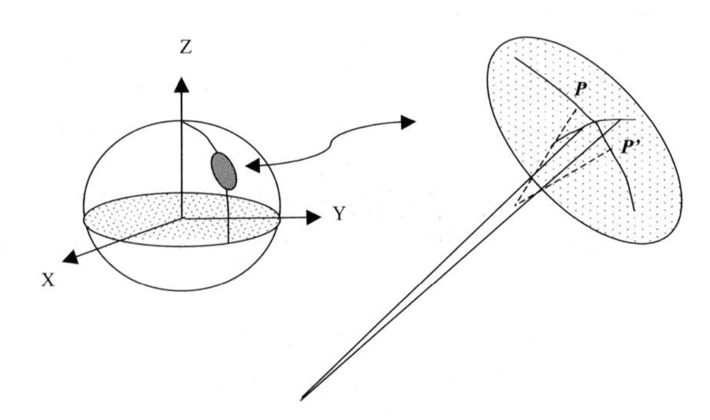

Fig. 2.12 Radius of curvature in meridian and prime vertical directions

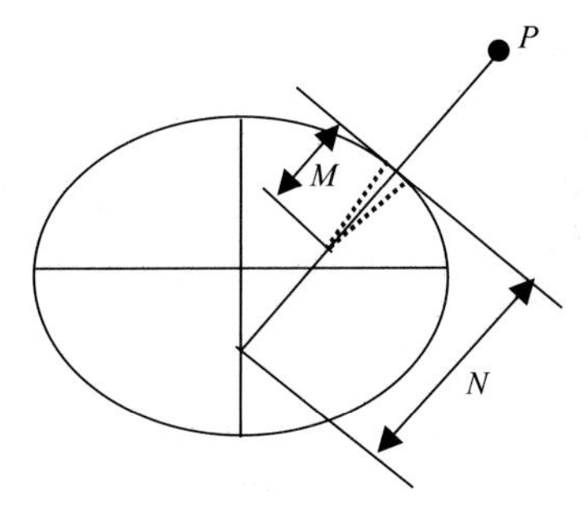

Fig. 2.13 Section view of the radius of curvature in meridian (M) and the prime vertical (N)

curvature of arc PP′ is the shortest. When PP′ are in the east-west direction, the radius of curvature of arc PP′ will be the longest. If PP′ is in a direction other than these two directions, the radius of curvature will be between the maximum and the minimum values. If P and P′ becomes closer and closer, then the line tangent to P is obtained. If this tangent line is in the north-south direction, then P has the radius of *curvature* "*in the meridian*". If the tangent line is in the east-west direction, then P has the radius of *curvature* "*in the prime vertical*". The name *"principal radii of curvature"* is given to the radius in these two particular directions, and they are usually denoted as M and N or ρ and ν. Figure 2.13 is the section view of M and N. The formulae for calculating M and N are as follows:

$$M = \frac{a(1-e^2)}{(1-e^2\sin^2\varphi_G)^{\frac{3}{2}}}$$
$$N = \frac{a}{\sqrt{(1-e^2\sin^2\varphi_G)}}.$$

2.4.2 Conversion between Geodetic and Cartesian Coordinates

For surveying and mapping applications, we prefer measurements on an ellipsoidal datum because this is the surface upon which we normally operate. There are other applications, particularly space and satellite navigation, that prefer the simple Cartesian coordinate frame fixed in space because they normally operate beyond Earth's surface. It is therefore necessary to convert from the geodetic coordinates in polar parameters $(\varphi_G, \lambda_G, h)$ to Cartesian coordinates using the following equations:

$$\mathrm{X} = (N+h)\cos\varphi_G\cos\lambda_G$$
$$\mathrm{Y} = (N+h)\cos\varphi_G\sin\lambda_G$$
$$\mathrm{Z} = \left[N(1-e^2)+h\right]\sin\varphi_G$$

where,

φ_G is the geodetic latitude
λ_G is the geodetic longitude
h is the height above (below) the ellipsoid surface.
N is the radius of curvature in prime vertical,
a is the semi-major axis of the ellipse used to generate the ellipsoid
e is the first eccentricity of the ellipse used to generate the ellipsoid.

The reverse process is more complicated and is beyond the scope of this book. The interested reader is referred to, for example, Schofield [1993].

2.5 Geodetic and Plane Coordinate Systems

The geodetic coordinates in latitudes and longitudes can accurately represent the position of a point on the surface of Earth, assuming, of course, that Earth is an ellipsoid. These geodetic coordinates can be used to precisely calculate length, area, and angle on an ellipsoid using rather complicated mathematics. For most applications, a plane coordinate system is more practical. A plane coordinate system, being a two-dimensional coordinate system, allows geometrical calculations to be done using plane geometry and trigonometry. The conversion between geodetic and plane coordinate systems is the study of map projections to be discussed in Chapter 4.

3 Transformation of Coordinates between Cartesian Systems

Xiao-Li Ding

3.1 Introduction

It is often necessary to convert spatial data from one reference system to another. For example, when one has determined the positions of some points in a local rectangular grid system (see *Chapter 2* for definitions of the various reference systems), the positions may be converted to a national grid system, or vice versa, for various reasons. The conversion of coordinates between different reference systems, when the relationships between the systems are known, is referred to as *coordinate transformation*.

This chapter will first look at the basic concept of coordinate transformation, and then introduce some of the commonly used coordinate transformation models. Finally, the methods for the determination of the relationship (transformation parameters) between reference systems for the transformation of coordinates will be discussed.

The transformation of coordinates discussed in this chapter will be limited to the transformation between Cartesian systems. For conversion of coordinates from one form (e.g., the geographical coordinates) into another (e.g., grid coordinates) within the same reference system, the reader is referred to Chapter 4.

3.2 Basic Concepts of Coordinate Transformations

The two basic problems of coordinate transformations are:

1) Transformation of coordinates from one system to another when the relationship between the two systems is known; and
2) Determination of the relationship between two reference systems when the coordinates of some points that are common in both of the two systems are known.

The relationship between two systems is usually described by certain parameters that are referred to as *transformation parameters*. Therefore, the second problem is often referred to as the determination of transformation parameters.

The first problem can be expressed using the following simple mathematical models:

$$\begin{aligned} X &= f_X(x, y, z) \\ Y &= f_Y(x, y, z) \\ Z &= f_Z(x, y, z), \end{aligned} \tag{3.1}$$

or

$$\begin{aligned} x &= f_x(X, Y, Z) \\ y &= f_y(X, Y, Z) \\ z &= f_z(X, Y, Z), \end{aligned} \tag{3.2}$$

where (x,y,z) and (X,Y,Z) are the three-dimensional coordinates of any point in the *x*-*y*-*z* and the *X*-*Y*-*Z* systems, respectively, and f_x, f_y, f_z, f_X, f_Y and f_Z are functions (transformation models) relating the two systems. Figure 3.1 shows 2 two-dimensional systems, *x*-*y* and *X*-*Y*, respectively.

Equation (3.1) transforms the coordinates of any point from the *x*-*y*-*z* system to the *X*-*Y*-*Z* system, and Equation (3.2) transforms the coordinates from the *X*-*Y*-*Z* system back to the *x*-*y*-*z* system. If Equation (3.1) is the forward transformation, then Equation (3.2) is the *reverse transformation*, and vice versa.

The second problem of coordinate transformation is to determine the functions f_x, f_y, f_z, f_X,f_Y and f_Z, or more specifically certain parameters in the functions, when the coordinates of some points (control points) are known in both of the two reference systems.

Figure 3.1 shows points *A*-*F* that are present in both *x*-*y* and *X*-*Y* systems. The reference systems are both two-dimensional right-handed Cartesian systems. The transformation model (functions relating the two systems) is

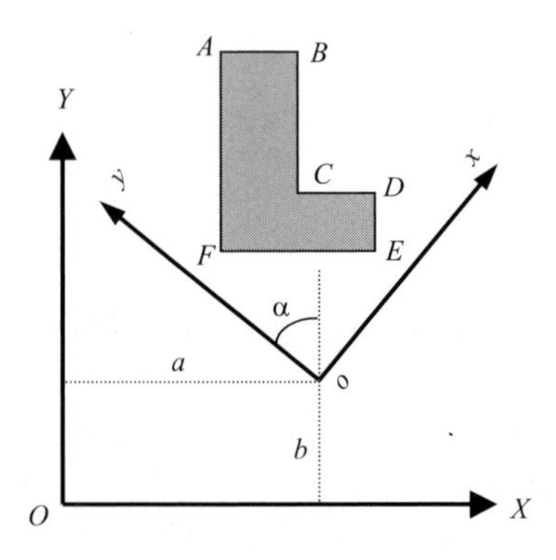

Fig. 3.1 Points in two different reference systems

Table 3.1 Coordinates of points in two coordinate systems (in metres)

Point	First System		Second System	
	x	y	X	Y
A	99.403	242.910	120.000	350.000
B	145.361	204.346	180.000	350.000
C	81.089	127.750	180.000	250.000
D	134.707	82.759	250.000	250.000
E	102.571	44.461	250.000	200.000
F	2.995	128.015	120.000	200.000

$$X = 200.000 + 1.0001(x \cos 40^\circ - y \sin 40^\circ)$$
$$Y = 100.000 + 1.0001(x \sin 40^\circ + y \cos 40^\circ)$$

or

$$x = \frac{1}{1.0001}[(X - 200.000) \cos 40^\circ + (Y - 100.000) \sin 40^\circ]$$
$$y = \frac{1}{1.0001}[-(X - 200.000) \sin 40^\circ + (Y - 100.000) \cos 40^\circ]$$

The coordinates of the points in the two systems are given in Table 3.1.

A coordinate transformation can often be interpreted as certain geometrical changes, typically *translation* (or *shift* of the origin), *rotation*, and *scale change*, through which one of the systems is made to overlap exactly the second. For example, for the two reference systems shown in Figure 3.1, these changes are:

- The translations of the origin are a = 200 m and b = 100 m;
- The rotation angle for the x-y axes to become parallel with the X-Y axes is $\alpha = 40^\circ$; and
- The scale factor between the two systems is 1.0001. The scale change (not shown in the diagram) can be considered to be caused by the different unit lengths in the axes of the two reference systems. Therefore, a one metre distance in one system becomes 1.0001 m in the other, and 10 km become 10.001 km, and so on.

3.3 Models for Coordinate Transformation

Different transformation models are used in practice. Each of these has properties that suit certain specific problems. This section will introduce some of the most commonly used transformation models. The properties of the models will also be given in brief to help the reader to differentiate between them and to choose the most appropriate ones to use when required.

3.3.1 Two-Dimensional Conformal Transformation

The conformal transformation is one of the most commonly used models. The model is also known as the *similarity* or *Helmert transformation.* The transformation shown in Figure 3.1 belongs to this type. The general two-dimensional conformal transformation model is:

$$\begin{aligned} X &= a + s(\cos\alpha)x - s(\sin\alpha)y \\ Y &= b + s(\sin\alpha)x + s(\cos\alpha)y. \end{aligned} \tag{3.3}$$

The parameters used in the above equations are (see Figure 3.1):

a, b – *translations* of the origin along the X and Y axes, respectively, or the coordinates of the origin of the x-y system (the old system) in the X-Y system (the new system);

α – the *rotation* of the axes of the x-y system to the X-Y system. Angle α is defined as positive when the rotation is clockwise, and as zero if the axes are parallel; and

s – the *scale factor* between the two systems. When the unit length of the new system is shorter than that of the old system, s is larger than 1.

Equation (3.3) will be derived later in section 3.4. This model transforms the coordinates of any point from the x-y system to the X-Y system. Both are right-handed Cartesian systems. Coefficients a, b, α, and s are the *transformation parameters.* These parameters define the relationship between the two systems.

Equation (3.3) is often written as:

$$\begin{aligned} X &= a + cx + dy \\ Y &= b - dx + cy, \end{aligned} \tag{3.4}$$

where a, b, c, and d represent another form of the four transformation parameters in the transformation model. Since in either Equation (3.3) or (3.4) four parameters define the relationship between the two coordinate systems completely, the model is a *four-parameter* transformation model.

In matrix notation, Equation (3.3) can be written as

$$\begin{pmatrix} X \\ Y \end{pmatrix} = \begin{pmatrix} a \\ b \end{pmatrix} + s\begin{pmatrix} \cos\alpha & -\sin\alpha \\ \sin\alpha & \cos\alpha \end{pmatrix}\begin{pmatrix} x \\ y \end{pmatrix} = \begin{pmatrix} a \\ b \end{pmatrix} + sR_\alpha\begin{pmatrix} x \\ y \end{pmatrix}, \tag{3.5}$$

where

$$R_\alpha = \begin{pmatrix} \cos\alpha & -\sin\alpha \\ \sin\alpha & \cos\alpha \end{pmatrix}$$

is the *rotation matrix* for angle α.

The conformal transformation model has the following properties:

a) The conformal transformation preserves the shapes of figures. For example, all the angles will be preserved. Additionally, straight lines will remain straight; squares will still be squares; and circles will still be circles when transformed from one system to another although the sizes and the orientation of the figures may change due to the scale change and the rotation.
b) The scale change between the two systems is the same everywhere and in any direction. This partly explains (a) above.
c) For R_α, the following is true,

$$\mathrm{R}_\alpha^{-1} = \mathrm{R}_\alpha^T = \mathrm{R}_{-\alpha}. \tag{3.6}$$

Using these properties, we know the reverse transformation of Equation (3.5) is,

$$\begin{pmatrix} \mathrm{x} \\ \mathrm{y} \end{pmatrix} = \frac{1}{\mathrm{s}} \begin{pmatrix} \mathrm{X} - \mathrm{a} \\ \mathrm{Y} - \mathrm{b} \end{pmatrix} \begin{pmatrix} \cos\alpha & \sin\alpha \\ -\sin\alpha & \cos\alpha \end{pmatrix}. \tag{3.7}$$

3.3.2 Two-Dimensional Affine Transformation Model

The two-dimensional affine transformation is a slight modification of the two-dimensional conformal transformation described in section 3.3.1. As we have already discussed in the previous section, the conformal transformation model assumes a uniform scale change everywhere and in all directions. The affine transformation model, however, uses two different scale factors for the x and the y directions, respectively. Besides, the affine transformation also allows changes in the orthogonality between the two axes of the coordinate systems (Figure 3.2).

The general form of the two-dimensional affine coordinate transformation model is

$$\begin{aligned} \mathrm{X} &= \mathrm{a} + \mathrm{s_x}(\cos\alpha)\mathrm{x} + \mathrm{s_y}(\sin\alpha\cos\beta - \cos\alpha\sin\beta)\mathrm{y} \\ \mathrm{Y} &= \mathrm{b} - \mathrm{s_x}(\sin\alpha)\mathrm{x} + \mathrm{s_y}(\sin\alpha\sin\beta + \cos\alpha\cos\beta)\mathrm{y}. \end{aligned} \tag{3.8}$$

Its six parameters are:

a, b – translations of the origin along the X and Y directions, respectively, or the coordinates of the origin of the x-y system (the old system) in the X-Y system (the new system);

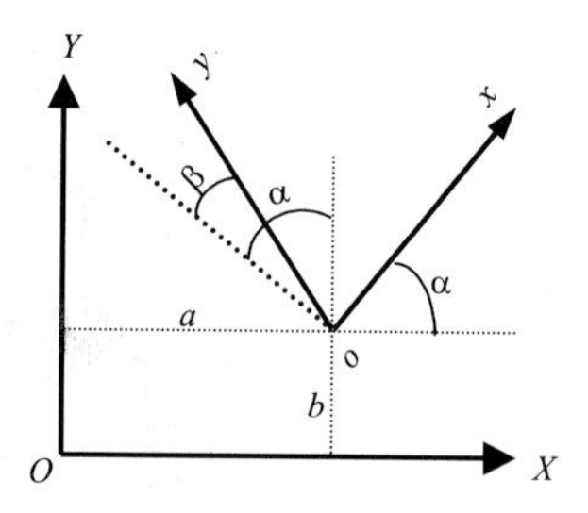

Fig. 3.2 Two-dimensional affine transformation

α – the rotation angle from the x axis to the X axis;
s_x, s_y – the scale factors in the x and y directions, respectively; and
β – the change in the orthogonality of the axes of the x-y system.

The model can also be written as

$$\begin{aligned} X &= a + cx + dy \\ Y &= b - ex + fy, \end{aligned} \tag{3.9}$$

where a, b, c, d, e, and f are another form of the six transformation parameters.

As seen in Equation (3.8), the affine transformation model uses two scale factors, s_x and s_y, allowing different scale changes in the x and the y directions. The model also allows changes in the angle between the coordinate axes. The affine transformation model, therefore, does not preserve shapes of figures, such that a square will usually become a parallelogram and a circle will become an ellipse. Despite this, straight lines will remain straight, and parallel lines will remain parallel in the transformation.

An example of affine transformation is in digitising maps. Existing maps may have shrunk or expanded differently along the two axes of the map. When the digitised map coordinates are transformed into a new reference system, this model is more appropriate than the conformal transformation model.

3.3.3 Two-Dimensional Projective Transformation

The general form of the two-dimensional projective transformation model is (Figure 3.3)

$$\begin{aligned} X &= \frac{ax + by + c}{gx + hy + 1} \\ Y &= \frac{dx + ey + f}{gx + hy + 1}, \end{aligned} \tag{3.10}$$

where a, b, c, d, e, f, g, and h are the eight transformation parameters defining the relationship between the two reference systems. The model

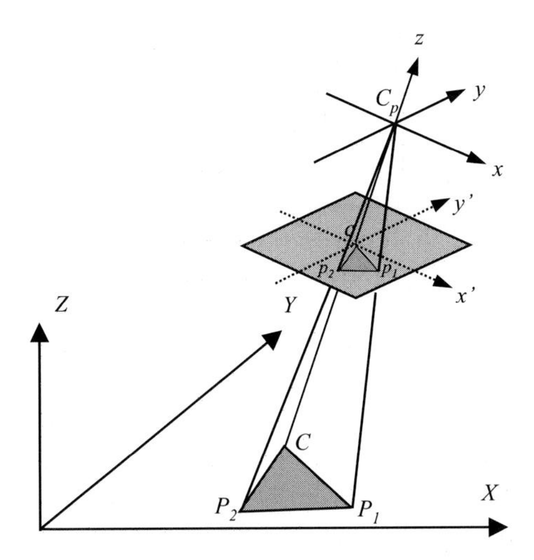

Fig. 3.3 Transformation of coordinates from x'-y' system to X-Y system through projection centre C_p

involves the transformation of one plane into another, through a point known as the projection centre (C_p in Figure 3.3).

It can be easily seen by comparing Equations (3.9) and (3.10), that if $g = h = 0$, the projective transformation becomes the affine transformation. Therefore, the affine transformation is a special case of the projective transformation.

The projective transformation does not preserve shapes of figures although it preserves straight lines.

3.3.4 Two-Dimensional Polynomial Transformation

Sometimes in practice, the distortion pattern involved in a transformation is unknown or the distortions are too complicated to be described using the models introduced above. In that case the polynomial transformation is useful.

The general two-dimensional polynomial transformation model is

$$\begin{aligned} X &= a_0 + a_1x + a_2y + a_3xy + a_4x^2y + a_5xy^2 + \cdots \\ Y &= b_0 + b_1x + b_2y + b_3xy + b_4x^2y + b_5xy^2 + \cdots, \end{aligned} \tag{3.11}$$

where coefficients $a_i, b_i (i = 0, 1, 2, \ldots)$ are the transformation parameters. The degree of the polynomial should be selected such that it can best model the relationship between the two systems. In general, the higher the degree, the better the model can describe the distortions. As a consequence of this, however, more control points are required to determine the transformation parameters (see 3.5 below).

If only the first three terms are selected, Equation (3.11) becomes

$$\begin{aligned} X &= a_0 + a_1 x + a_2 y \\ Y &= b_0 + b_1 x + b_2 y. \end{aligned} \tag{3.12}$$

Comparing this equation to Equation (3.9), we can see that the affine transformation is a special case of the polynomial model. Further, if in Equation (3.12) we let

$$\begin{aligned} a_1 &= b_2 \\ a_2 &= -b_1, \end{aligned} \tag{3.13}$$

then Equation (3.12) becomes the conformal transformation model given in Equation (3.4). Therefore, the conformal transformation model is a special case of both the affine and the polynomial transformation models.

3.3.5 Three-Dimensional Conformal Transformation Model

The general form of the three-dimensional conformal transformation model is,

$$\begin{pmatrix} X \\ Y \\ Z \end{pmatrix} = \begin{pmatrix} x_0 \\ y_0 \\ z_0 \end{pmatrix} + sR(\omega_1, \omega_2, \omega_3) \begin{pmatrix} x \\ y \\ z \end{pmatrix}, \tag{3.14}$$

where (x,y,z) are the coordinates of any point in the x-y-z system and (X,Y,Z) are the coordinates of the same point in the X-Y-Z system. (x_0, y_0, z_0) are the translations or the shift of the origin in the directions of the three axis X, Y and Z, respectively. They are also the coordinates of the origin of the x-y-z system in the X-Y-Z system. s is the scale factor and $R(\omega_1, \omega_1, \omega_1)$ is the rotation matrix,

$$R(\omega_1, \omega_2, \omega_3) = R_1(\omega_1) R_2(\omega_2) R_3(\omega_3), \tag{3.15}$$

where

$$R_1(\omega_1) = \begin{pmatrix} 1 & 0 & 0 \\ 0 & \cos\omega_1 & \sin\omega_1 \\ 0 & -\sin\omega_1 & \cos\omega_1 \end{pmatrix}, \quad R_2(\omega_2) = \begin{pmatrix} \cos\omega_2 & 0 & -\sin\omega_2 \\ 0 & 1 & 0 \\ \sin\omega_2 & 0 & \cos\omega_2 \end{pmatrix},$$

$$R_3(\omega_3) = \begin{pmatrix} \cos\omega_3 & \sin\omega_3 & 0 \\ -\sin\omega_3 & \cos\omega_3 & 0 \\ 0 & 0 & 1 \end{pmatrix} \tag{3.16}$$

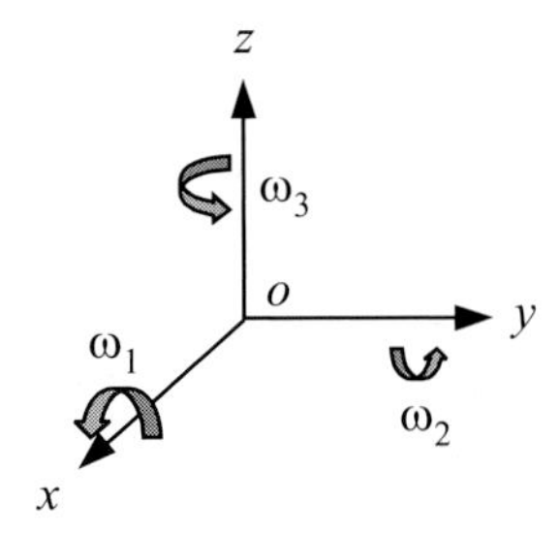

Fig. 3.4 Rotation angles in three-dimensional conformal transformation

are the fundamental rotation matrices describing the rotations around the x, y, and z axes respectively (Figure 3.4). The angles are defined as positive when the axes of the old system are rotated clockwise to the directions of the new system. Similar to Equation (3.6), the following relationships are true for the matrices

$$R_k(\omega_k)^{-1} = R_k(\omega_k)^T = R_k(-\omega_k). \qquad (3.17)$$

These can be used in the computation of the reverse transformation.

The shapes of figures are preserved in the three-dimensional conformal transformation. For example, a circle or a sphere will be maintained as a circle or a sphere after the transformation, and a cube will be maintained as a cube although the sizes and the orientations of the figures may change.

The reverse transformation for the three dimensional model is

$$\begin{pmatrix} x \\ y \\ z \end{pmatrix} = \frac{1}{s} R_3(-\omega_3) R_2(-\omega_2) R_1(-\omega_1) \left[\begin{pmatrix} X \\ Y \\ Z \end{pmatrix} - \begin{pmatrix} x_0 \\ y_0 \\ z_0 \end{pmatrix} \right]. \qquad (3.18)$$

When the scale factor s is close to 1, it is often expressed in the following form:

$$s = 1 + k \qquad (3.19)$$

where k is the actual *change in scale*. When k is small, it is commonly given in *parts per million* (ppm). For example, if $s = 1.000038$, then $k = 0.000038 = 38$ ppm.

As seven parameters, $x_0, y_0, z_0, s, \omega_1, \omega_2$, and ω_3, are used in Equation (3.14), this is a seven-parameter transformation model.

The three-dimensional conformal transformation model has been widely used in recent years to transform coordinates between geodetic systems. In the past, most nations (or regions) used their own national reference systems (geodetic datums). However, since the GPS (Global Positioning System, see Chapter 7) has been used commonly in recent years and the GPS uses a reference system called WGS84 (World Geodetic System 1984) that is different from the systems used before by most nations, it is often necessary to transform coordinates between the national (or regional) systems and the

WGS84. For example, the model has been used in Australia to transform coordinates between the Australian national reference system AGD84 (Australian Geodetic Datum 1984) to WGS84. The transformation parameters that have been in use for the transformation from WGS84 to AGD84 are [Higgins, 1987]:

x_0	116.00 (m)	±2.3 (m)
y_0	50.47 (m)	± 2.3 (m)
z_0	−141.69 (m)	±2.5 (m)
k	−0.098 (ppm)	±0.07 (ppm)
ω_1	0.230 (seconds)	±0.04 (seconds)
ω_2	0.390 (seconds)	±0.04 (seconds)
ω_0	0.344 (seconds)	±0.04 (seconds).

3.4 Derivation of Transformation Models

In section 3.3, the various transformation models were given without derivations. To get further insight into these models, the two-dimensional conformal transformation model given in Equation (3.3) will be derived below to illustrate how a transformation model can be derived considering the various geometrical changes.

The derivation of the other transformation models can be carried out in a similar fashion.

Translation of Coordinate Axes or Shift of Origin

If there is only a shift of origin between the two coordinate systems (from o to O in Figure 3.5a), the coordinates of any point P in the new system become,

$$\begin{aligned} X_P &= a + x_P \\ Y_P &= b + y_P \end{aligned} \tag{3.20}$$

where a and b are the coordinates of the origin o in the X-Y system.

Rotation of Coordinate Axes

If there is an angle α between the axes of the x-y system and the axes of the X-Y system (Figure 3.5b), the coordinates of point P in the two systems are:

$$\begin{aligned} X_P &= r\sin(\phi - \alpha) \\ Y_P &= r\cos(\phi - \alpha) \end{aligned} \tag{3.21}$$

$$\begin{aligned} x_P &= r\sin\phi \\ y_P &= r\cos\phi. \end{aligned} \tag{3.22}$$

From the following trigonometric identities,

$$\begin{aligned} \sin(\phi - \alpha) &= \sin\phi\cos\alpha - \cos\phi\sin\alpha \\ \cos(\phi - \alpha) &= \cos\phi\cos\alpha + \sin\phi\sin\alpha \end{aligned} \tag{3.23}$$

Equation (3.21) becomes

$$\begin{aligned} X_P &= r\sin\phi\cos\alpha - r\cos\phi\sin\alpha \\ Y_P &= r\cos\phi\cos\alpha + r\sin\phi\sin\alpha. \end{aligned} \tag{3.24}$$

Substituting Equation (3.22) in the above, one gets

$$\begin{aligned} X_P &= x_P\cos\alpha - y_P\sin\alpha \\ Y_P &= y_P\cos\alpha + x_P\sin\alpha. \end{aligned} \tag{3.25}$$

Using matrix notation, the above becomes,

$$\begin{pmatrix} X_P \\ Y_P \end{pmatrix} = \begin{pmatrix} \cos\alpha & -\sin\alpha \\ \sin\alpha & \cos\alpha \end{pmatrix} \begin{pmatrix} x_P \\ y_P \end{pmatrix} = R_\alpha \begin{pmatrix} x_P \\ y_P \end{pmatrix} \tag{3.26}$$

The matrix

$$R_\alpha = \begin{pmatrix} \cos\alpha & -\sin\alpha \\ \sin\alpha & \cos\alpha \end{pmatrix} \tag{3.27}$$

is the rotation matrix. The rotation angle α is defined as positive when the axes of the old system (x-y) are rotated clockwise to the new system (X-Y). It is easy to prove that if the old system is rotated anti-clockwise (the α angle is defined as negative) (Figure 3.5), the rotation matrix becomes

$$R_\alpha = \begin{pmatrix} \cos\alpha & \sin\alpha \\ -\sin\alpha & \cos\alpha \end{pmatrix}. \tag{3.28}$$

Scale Change

Scale change results from the different unit lengths used in the two reference systems. For example, if the old system uses the Imperial system and the new system uses the metric system, then all the coordinates in the old system need to be scaled by 1/3.2808 to change to the coordinates in the new system. For the reverse operation, the scale factor is 3.2808.

In general, the scale change in coordinate transformation can be expressed as,

$$\begin{aligned} X_P &= s x_P \\ Y_P &= s y_P \end{aligned} \tag{3.29}$$

or in matrix notation as,

$$\begin{pmatrix} X_P \\ Y_P \end{pmatrix} = s \begin{pmatrix} x_P \\ y_P \end{pmatrix} \tag{3.30}$$

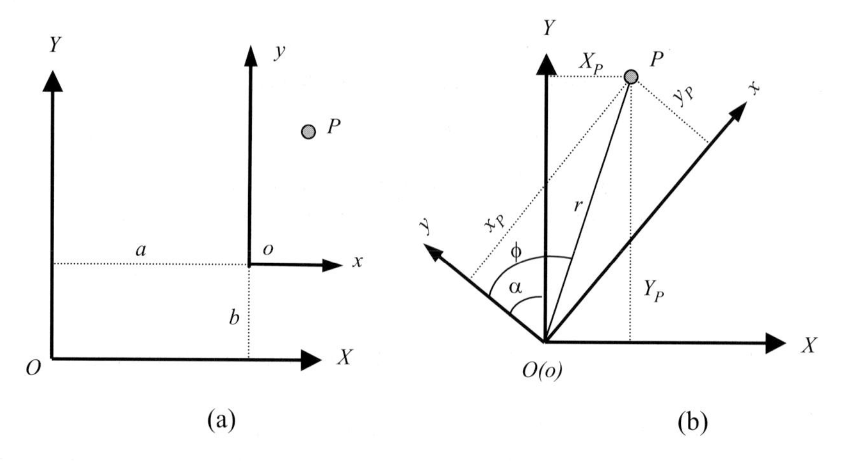

Fig. 3.5 Translation and rotation in coordinate transformation

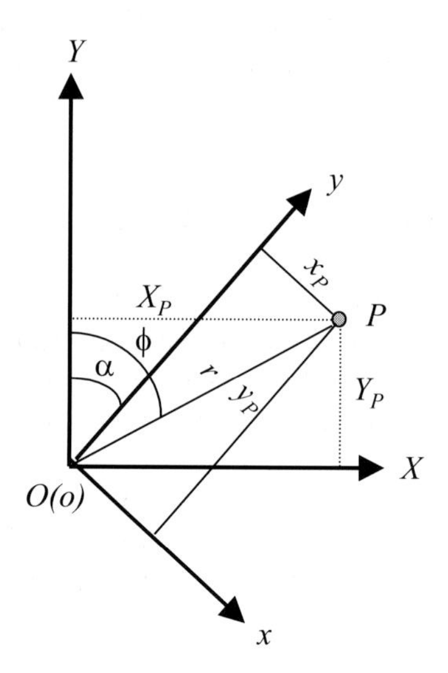

Fig. 3.6 Anti-clockwise rotation

where s is the scale factor to convert coordinates from the old to the new system.

Combination of Translation, Rotation, and Scale Change

Considering all three components of transformation, i.e., translation, rotation, and scale change, the following general two-dimensional conformal transformation model can be obtained by combining Equations (3.20), (3.26), and (3.30),

$$\begin{pmatrix} X_P \\ Y_P \end{pmatrix} = \begin{pmatrix} a \\ b \end{pmatrix} + s \begin{pmatrix} \cos\alpha & -\sin\alpha \\ \sin\alpha & \cos\alpha \end{pmatrix} \begin{pmatrix} x_P \\ y_P \end{pmatrix} = \begin{pmatrix} a \\ b \end{pmatrix} + s\mathbf{R}_\alpha \begin{pmatrix} x_P \\ y_P \end{pmatrix} \tag{3.31}$$

or

$$\begin{aligned} X_P &= a + s\cos\alpha x_P - s\sin\alpha y_P \\ Y_P &= b + s\sin\alpha x_P + s\cos\alpha y_P. \end{aligned} \tag{3.32}$$

In the above Equation (3.32), subscript p can be any point in the two systems. Equation (3.3) is obtained when the subscript is omitted.

3.5 Determination of Transformation Parameters

We have seen in Section 3.4 that a set of transformation parameters should be known before a transformation can be performed, no matter which transformation model is used. These parameters define the relationship between the two coordinate systems in question. In practice, the parameters are usually determined based on the coordinates of a number of points (known as control points) that are common in both systems.

3.5.1 Method of Direct Solution

The coordinates of the control points can be substituted in the chosen transformation model to form some simultaneous equations. When the number of equations thus formed is the same as the number of unknown transformation parameters, the equations can be solved for the parameters.

For example, for the case shown in Figure 3.1, if the coordinates of points A and E are known, and the two-dimensional conformal transformation model (3.4) is selected, the following equations can be formed:

$$\text{For point A}: \begin{cases} 120.000 = a + 99.403c + 242.910d \\ 350.000 = b - 99.403d + 242.910c \end{cases}$$

$$\text{For point E}: \begin{cases} 250.000 = a + 102.571c + 44.461d \\ 200.000 = b - 102.571d + 44.461c \end{cases}$$

Solving the four simultaneous linear equations for the four unknown parameters a, b, c and d, we can get,

$$\begin{aligned} a &= 199.9996 \\ b &= 99.9996 \\ c &= 0.76612403 \\ d &= -0.64284990. \end{aligned}$$

According to Equations (3.3) and (3.4), we can see that

$$\begin{aligned} \mathrm{c} &= \mathrm{s}(\cos\alpha) \\ \mathrm{d} &= -\mathrm{s}(\sin\alpha). \end{aligned} \tag{3.33}$$

Therefore

$$\begin{aligned} \alpha &= -\arctan\left(\frac{d}{c}\right) = 39°59'59.3'' \\ s &= \sqrt{c^2 + d^2} = 1.000101 \end{aligned} \tag{3.34}$$

a, b, α and s are the translations, the rotation, and the scale factor between the two systems. The small differences between the results and the known values (see section 3.2):

$$\begin{aligned} a &= 200\,\mathrm{m} \\ b &= 100\,\mathrm{m} \\ \alpha &= 40° \\ s &= 1.0001 \end{aligned}$$

are caused by the significant figures used in the computations.

It can be seen from the above example that two equations can be set up using the coordinates of each control point, therefore, at least two control points are required to solve for the four parameters. As a general rule, for any of the transformation models given in section 3.3, the number of equations required always should be the same as the number of unknown parameters. If the number of equations is less than the number of parameters, no solution can be obtained. On the other hand, if the number of equations is more than the number of parameters, multiple solutions result. The two situations are referred to as the *under determined* and the *over determined* problems.

Remember that each control point has 2 coordinates in a two-dimensional model, and 3 coordinates in a three-dimensional model, if all the coordinates of a point are used in the solution. According to this rule, we can determine the number of points required for all the different models as given in Table 3.2.

As Equation (3.10) is non-linear, linearisation is required when solving the transformation parameters for the model. Interested reader may consult, for example, Wolf and Ghilani (1997) on the topic.

Since the control points used for solving for the transformation parameters usually contain errors, different outcomes may result when different control points are used. To demonstrate this point, errors with a standard deviation of ± 5 mm have been simulated and added to the coordinates in the x-y system in Table 3.1 to form the following new coordinates (Table 3.3).

Table 3.2 The minimally required number of points for solving transformation parameters

Model	Number of control points required
2-D conformal	2
2-D affine	3
2-D projective	4
2-D polynomial	Depending on the degree of the polynomial, e.g., 1st degree: 3, 2nd degree: 6.
3-D conformal	2 plus one coordinate from a 3rd point

Table 3.3 Coordinates with simulated errors added

Point	First System		Second System	
	x	y	X	Y
A	99.391	242.921	120.000	350.000
B	145.373	204.348	180.000	350.000
C	81.087	127.746	180.000	250.000
D	134.696	82.752	250.000	250.000
E	102.575	44.466	250.000	200.000
F	3.001	128.011	120.000	200.000

When points A and E are used again to calculate the transformation parameters, the results obtained become:

$$
\begin{aligned}
\mathrm{a} &= 199.993\,\mathrm{m}\\
\mathrm{b} &= 100.000\,\mathrm{m}\\
\alpha &= 39^\circ 59' 42''\\
\mathrm{s} &= 1.000072.
\end{aligned}
$$

When points A and F instead of A and E are used for the calculation, the following parameters are obtained,

$$
\begin{aligned}
\mathrm{a} &= 199.978\,\mathrm{m}\\
\mathrm{b} &= 99.985\,\mathrm{m}\\
\alpha &= 39^\circ 59' 27''\\
\mathrm{s} &= 1.000109.
\end{aligned}
$$

Clearly it can be seen from the example that the calculated transformation parameters are affected by the errors in the coordinates used in the calculation.

A common way to reduce the effect of errors is to use more control points than the minimum required. This has the same effect as taking several measurements of a distance between two points and then accepting the mean of the measurements to reduce the effect of measurement errors. When more points are used, however, we face the problem of over determination and multiple and inconsistent solutions may result. This problem has been seen in the above example.

3.5.2 The Method of Least Squares

Least squares is an estimation method that has been widely used in many fields including surveying and mapping. The method can produce a unique solution from an over determined equation system, and is therefore ideal for the problem of determining transformation parameters when more than the minimum number of control points is used. The solution obtained using the least-squares method has certain desirable properties. For example, when the errors in the observations are normally distributed, the solution is the most probable one.

Aside from the transformation parameters, as a by-product, the least-squares solution can also produce the so-called *least-squares residuals*. Most software packages output these. The residuals reflect how well the control points in the two systems and the chosen transformation model fit to each other. Large residuals are usually due to:

(a) large errors in the coordinates of one or more of the control points; and
(b) an inappropriately chosen transformation model.

Examples of (a) include digitisation errors and mistakes in entering coordinates in the computer. The situation of (b) can result if, for example, the affine transformation model can best describe the relationship between two systems but the conformal transformation model is used.

Cautions should be excised while examining the residuals. Small residuals do not always mean good solutions. For example, if only the minimally required number of control points is used, all the residuals will be zero. Other factors such as the weighting of the coordinates in the solution also affect the residuals.

A detailed treatment of the least-squares theories is beyond the scope of this book. Interested readers are referred to Harvey [1990], Mikhail [1976] and Wolf and Ghilani [1997] for further details.

3.5.3 Guidelines for Practical Computations

Here are some general remarks on the determination of transformation parameters.

1) In general, the more control points used in the solution, the more accurate the results should be.

2) The spatial distribution of the control points used for solving for the transformation parameters is also important. It is preferable to use evenly distributed points and to ensure that the control points cover the entire area affected by the transformation. In another word, the points to be transformed should preferably be bounded by the control points rather than falling outside of them.
   For example, for the points in Figure 3.1, it is more advantageous to use A, B, D, E, and F as control points to solve for transformation parameters to transform point C than using points B, C, D and E as control points to transform points A and F.
3) When the least-squares method is used to solve for transformation parameters, dissimilar solutions may be obtained under the following different assumptions:
   - Only the coordinates of the control points in the first (old) system have errors;
   - Only the coordinates of the control points in the second (new) system have errors; and
   - The coordinates of the control points in both systems have errors.

   The solutions under the first two assumptions apply to situations in which the coordinates in one of the systems are much more accurate than those in the other. Therefore the errors in the coordinates of one of the systems can be neglected. The solution obtained under the third assumption is the most rigorous, but the solution is more complicated. In practice, one is often limited by the software available for the computation. Most commercial software packages use the approximate solutions and do not give the user much choice.
4) It is advisable to check the control points for gross errors when carrying out the computation. The errors can come from field surveys, computations, transcriptions, or typing. If unnoticed, gross errors can significantly affect the solved parameters. A method for checking the errors is to leave some control points out of the solution, but to transform these points using the calculated parameters. A comparison of the transformed and the known coordinates of the points can give a good indication of the quality of the transformation model.

4 Map Projections

Yuk-Cheung Lee

4.1 Introduction

Ever since we realised that the earth is not flat, we have been faced with the challenge of depicting features on earth on a flat surface as accurately as possible, thus the study of map projections. It is impossible for any map projection to produce a geometrically accurate map. The development of many different kinds of map projections throughout the ages has tried to make maps less distorted in one way or another.

Some of us have come to learn to live with these distortions to the extent that the uninitiated might not even know that maps are geometrically distorted. Indeed, it is true that for large scale applications, the implications of these geometric distortions due to map projections are negligible because the part of the curved earth covered by each one of these maps is practically flat. Traditionally, we need to understand map projections in order to be aware of the magnitude of the geometrical errors from measurements made on maps. After all, maps were, and to some extent still are, important analytical tools.

With the advent of digital maps, the role of paper maps as analytical tools has been reduced. At the same time, the visibility of map projections in digital map databases has increased. This is because we need to store map projection information as part of the database and we need to know about it in various operations.

When digital mapping software was first developed in the early 1960s, existing maps were the most cost effective source of data, and they naturally became the main data source. Software developed during that era stored data in the coordinate system of the map projection. These are known as *ground coordinates*, usually in units of metres, which are based on a Cartesian system with orthogonal axes in two dimensions. Today, maps (or hardcopy plots of one form or another) still remain as important input documents. The use of *geographic coordinates* in latitudes and longitudes to store data in GIS software is only beginning to emerge.

Many GIS operations require geographic coordinates. In the simplest case, we need to read the geographic coordinates of locations on a map. For more accurate calculations of distance, area, and angle, based on a curved surface of the earth instead of a flat approximation of it, GIS software needs to convert the stored coordinates in the Cartesian system back to the geographic system. To perform the conversion, the software needs to know the map projection used and all its necessary parameters. The software will request the user to enter such information before digitising of the map can begin.

A GIS database is designed for particular applications that might need the special characteristics of a certain map projection. The map projection required might not be the same as that of the input map. A conversion would be required and, to perform the calculations, the conversion software needs to know both the map projections and their parameters. The same applies when we export a digital map from one GIS system to another.

Integrating data in different map projections might or might not require map projection conversions depending on the level of accuracy required. If the area covered is small, we can assume that the surface is flat and there would be no need for map projections.

This chapter explains the basic concepts of map projections, the characteristics of the main types, the parameters required to define a map projection, and their use in a digital environment. We will give special consideration to the Universal Transverse Mercator (UTM) projection, the most popular projection for topographic maps.

4.2 The Shape of the Earth

A more detail description of the earth's shape and the coordinate systems used for referencing positions are given in Chapter 2 of this book. We will reiterate here that the exact shape of the earth is too complex for mathematical representation. Consequently, geometrical approximations to the earth using either a sphere or an ellipsoid are often used. For small-scale atlas mapping, the sphere is considered accurate enough. Calculations on an ellipsoid is much more complex and is beyond the scope of this book. Hence we have assumed a spherical earth in this chapter.

If a spherical or an ellipsoidal surface were used to display a map, then the problem of transferring features on earth to it given their geographic coordinates would have been trivial. The only difference between the two systems would have been the scale. Unless one is using a globe-like display surface, however, the matter of converting coordinates on an ellipsoidal earth to a very differently shaped display surface can be complicated.

4.3 Map Projections as Mathematical Transformations

Flat maps are more common than globes because they are easier to carry around and to work on. Hence the complicated task of converting features from an ellipsoidal surface to a flat one (*planar surface*) to produce useful maps cannot be avoided. If a sphere were to be used the conversion would be simpler but it would still involve a lot of calculations.

There is, however, a simple analogue method of producing some of the map projections using a transparent globe (the *reference body*) with features on it, a light source (the *projection centre*), and a flat sheet of translucent paper (the *mapping plane*). After placing the flat paper on the globe and the light source at the proper location, one can trace features projected from the globe onto the paper, hence the name map projections. Figure 4.1 illustrates a map projection with a light source at the centre of the spherical earth and the flat mapping plane touching the earth at the North Pole. In this projection, the parallels are projected as concentric circles that are longer than their true length. This distortion in length increases towards the edge of the map and reaches infinity at the equator, thus making it impractical to map close to it. The meridians are straight lines radiating from the North Pole. The lengths of the meridians are not correct because the scale along them increases towards infinity in the direction of the equator. A full circle around the North Pole covers 360 degrees on earth as well as on this map, meaning that all angular measurements from this point are preserved correctly.

A map projection is therefore a conversion, or a *mathematical transformation*, between two coordinate systems: the geographic system in latitude and longitude (ϕ, λ) and a system based on the plane surface. There are two popular choices in the latter case: one is the *Cartesian system*, and the other is the *polar system*. The Cartesian system yields (X, Y) coordinates (Figure 4.2), whereas the polar system yields (Range, Angle).

The Cartesian and polar systems are convertible – one can digitize a polar coordinate grid drawn on a flat sheet of paper and reproduce it

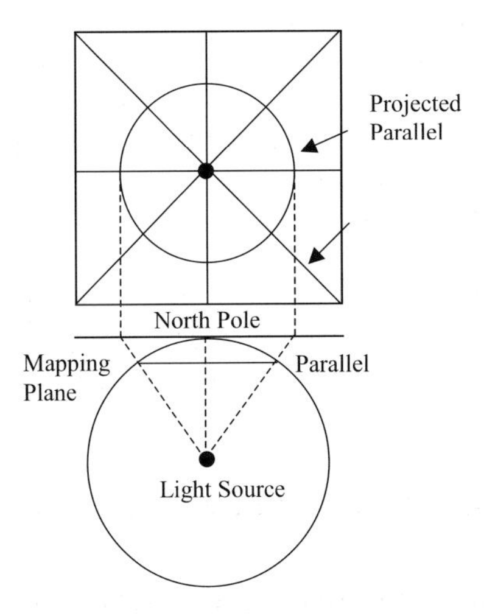

Fig. 4.1 A map projection by light

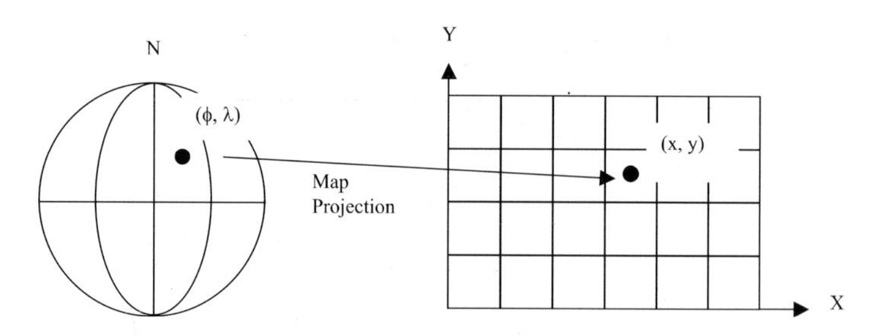

Fig. 4.2 Map projection as a transformation of coordinate systems

exactly (if you use enough points) on another paper by plotting the points. Hence, we can simplify the case and regard a map projection as a transformation function between the (ϕ, λ) system and the (X, Y) system. In other words,

$$f(\phi, \lambda) \rightarrow (X, Y), \text{ or}$$
$$f(\phi, \lambda) \rightarrow (Y, X)$$

if it is important to match latitude with Y instead of X coordinates.

A map projection is expressed as a set of equations that could take a point expressed in (ϕ, λ) and convert it to its equivalent position on a flat map in (X, Y). In the ideal (but impossible) case, all geometric properties of features in the (ϕ, λ) system are preserved in the (X, Y) system. Of all the geometric properties, we are particularly interested in length, area, and angle. Shape is also an important factor, but preserving length (meaning a constant scale in all directions) and angle automatically preserves shape. The sad fact is that no map projection can preserve length, area, and angle at the same time. In other words, one way or another all flat maps are wrong geometrically.

4.4 Geometric Distortions on Maps

Among the three main types of geometric distortions, length and area distortions are easiest to understand. Length and area distortions mean that length and area measured on a map are not correct. By measurement, we mean the use of plane geometry and trigonometry involving Cartesian coordinates to perform the calculations.

We have to be careful here concerning the measurement of point locations on a map. Let us suppose the graticule of meridians and latitudes is reproduced on a flat map through a map projection. We can measure the location of a point on the Cartesian system yielding a reading in X and Y, such as (370 km, 310 m) in Figure 4.3. Alternatively, we can interpolate the location in latitude and longitude using the geographic graticule that could

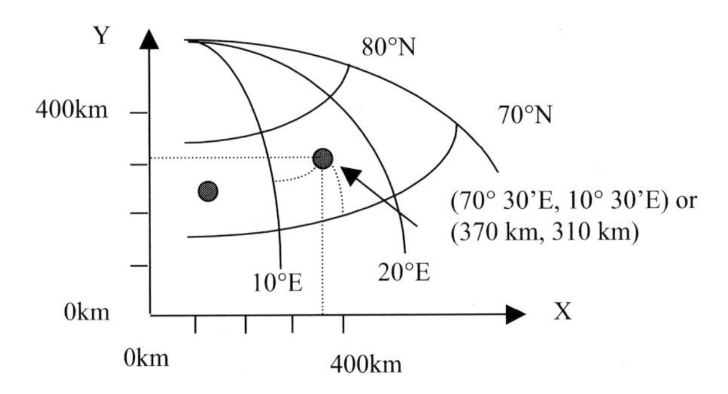

Fig. 4.3 Interpolating Cartesian and geographic coordinates

consist of curved lines, such as (70°30′E, 10°30′E) in Figure 4.3. Interpolation in this case will be over a curved network instead of the rectangular grid for the Cartesian system. Using this method, one can actually get a very good reading of the location of the point in geographic coordinates. We will bring up this point again later to illustrate that it is possible to obtain exact geographic coordinates from flat maps. However, this cannot be regarded as measurement in the Cartesian system because a different set of geometrical rules is used.

The reading of geographic coordinates with high accuracy is perhaps possible on flat maps using non-linear interpolation on the geographic graticule. The measurements of length, area, and angle are very different matters. In the case of length, unless geodetic calculations on an ellipsoid are used, direct measurement on the map would not yield correct results no matter how accurately one can read latitudes and longitudes along the line one needs to compute length. This is because direct measurements are done using rules of plane trigonometry.

Angular distortions are more difficult to understand, and we will elaborate on that here using an example involving both angle and length. Planar and ellipsoidal surfaces are geometrically very different. If the earth were flat, the rules of plane geometry and trigonometry would apply. In particular, the shortest distance between two points is a straight line and that straight line has a constant bearing, which is an angular measurement from north. North in this case is assumed to be at the top of the flat map, and all lines pointing north are parallel. On a spherical surface, let alone an ellipsoidal one, these two characteristics take on different meanings. To start with, the concept of a straight line does not quite apply to a curved surface although the concept of the shortest distance between two points does.

It is well known that the shortest distance between two points on a sphere (a *geodesic*) follows a *great circle*, which is a circle centred at the centre of the spherical earth. It should be noted, however, that such a path changes in azimuth except along certain directions, such as when the path follows the equator or a meridian. Note that azimuth on a sphere or an ellipsoid

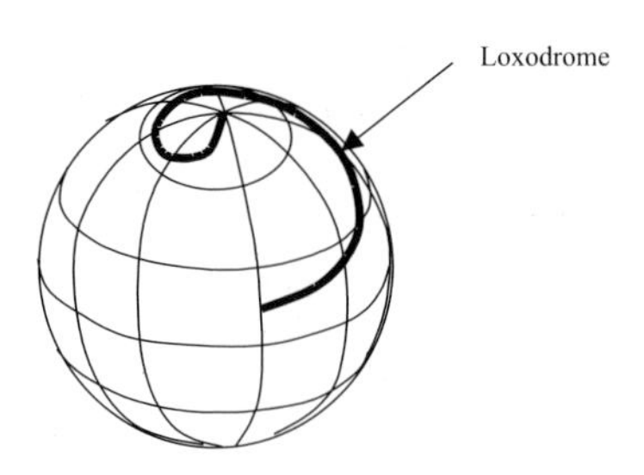

Fig. 4.4 Loxodrome

corresponds to bearing on a plane, and it is an angular measurement from the meridian north.

A path of constant azimuth on the sphere, called a *loxodrome* (or *rhumb line* in navigation), follows a complex curve called a spherical helix that spirals up towards the North Pole or down towards the South Pole (Figure 4.4). Again, loxodromes along the equator, any meridian, and any parallel of latitude are exceptions, and they form circular paths instead. In general, a line of constant azimuth is not always the shortest line between two points on a sphere.

This difference in the definition of a straight line shows a problem in converting features from a sphere or an ellipsoid to a planar surface. We will use this to illustrate one of the reasons for developing different kinds of map projections.

Other than a display of spatial features, maps also serve as analytical tools. Many maps were designed to make certain kinds of geometric measurements easier. This function has declined in importance with the advance of digital maps, where measurements can be done directly from the stored coordinates. Navigators were some of the earliest users to require rather accurate measurements in distance and bearing from charts. Take the problem of finding the shortest route between two points as an example. On a flat map it would be nice if a straight line between two points accurately represents the shortest route on a sphere. With such a chart, a navigator interested in the shortest route could simply join the origin and the destination by a straight line and measure the distance.

Indeed, a map projection that could "almost" produce such a chart had been developed around 500 years B.C. and is called the *Gnomonic* (or *Central*) projection. This is the projection we have introduced in section 4.3 as shown in Figure 4.1 and again in Figure 4.5a. All great circle routes on this projection are straight lines, although the lines are not necessarily of correct length. The reason why great circles are straight lines on this projection is because great circles pass through the centre of the earth, which is the centre of this projection as shown in Figure 4.1. The intersection between this circle (a plane) and the mapping plane is always a straight line.

This straight line, as shown in Figure 4.5a, does not follow a constant azimuth because it intersects the meridians at different angles. This causes a

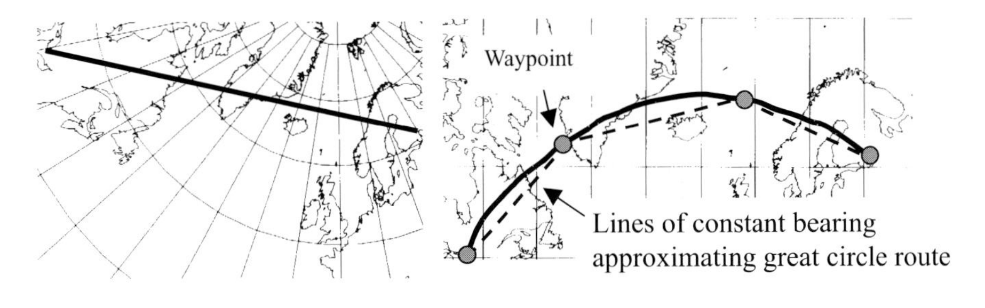

a) Gnomonic projection b) Mercator projection

Fig. 4.5 Great circle route on Gnomonic and Mercator projections

problem with navigators who prefer to follow a route with a constant azimuth because to continuously change the heading of a ship is an extremely tedious task. Hence, there is a need for another kind of map that can show a line of constant azimuth as a straight line. Gerardus Mercator in 1569 developed such a projection, and it has since been called the *Mercator projection* (Figure 4.5b).

Graticules on a Mercator projection form rectangular grids, very much like the grids in a Cartesian system. With this in mind, it is not hard to see why straight lines in a Mercator projection follow a constant azimuth. With a chart in the Mercator projection, the navigator can join two points with a straight line and then read off the azimuth from the map to set the compass bearing. As we know now, this line of constant azimuth is not necessarily the shortest route between two points.

Another advantage of the Mercator projection offered by its rectangular graticule is that geographic coordinates can be interpolated from it with very high accuracy using linear interpolation. It is easy to mistake the Mercator projection as one that observes rules of plane geometry and trigonometry because of its rectangular graticule. While this projection allows the reading of geographic coordinates by the rules of Cartesian geometry, it does not produce correct distances (among other geometric calculations) through rules of Cartesian geometry.

The Gnomonic and Mercator projections illustrate the difficulty in preserving angular measurements and length, two of the geometric properties, on a flat map. In the case of the Gnomonic projection, a shortest line along a great circle not necessarily constant in azimuth becomes a straight line with incorrect length on a flat map. In the case of the Mercator projection, a spiral line following a constant azimuth becomes a straight line also of incorrect length.

The two projections also illustrate the common practice of sacrificing some of the geometric properties in exchange for special characteristics on a map to serve the particular needs of an application. For navigation, both the Gnomonic and Mercator projections are useful in different ways. The usual practice is to first set off a shortest route (a great circle route) on a chart in the Gnomonic projection by joining two places with a straight line (Figure 4.5a). Then this is approximated by a series of segments of constant azimuth (Figure 4.5b). This is done by first selecting appropriate points for changes in compass bearing, called the *waypoints* by mariners, and then transferring these points to another chart in the Mercator projection. Finally, the way points on the Mercator projection will be joined by straight lines and compass bearings measured from them. Mariners will eventually use the approximation to the great circle route on the Mercator projection for navigation, changing heading only at waypoints.

We have in this discussion introduced the concepts of azimuth and bearing. We have also mentioned angle, which we will define more precisely here. An angle is measured from an observation point to two other points. An azimuth is a special kind of angle in that the line joining the observation point and one of the points is along a meridian or direction to true north. A description of how to perform angular measurements on the ground using surveying instruments is given in Chapter 6. On a map projection, that surveyed angle can be graphically reconstructed in the following way. From the observation point to each of the points is a great circle route, which could be a curved line on a map. Take the tangent of the two great circle routes at the observation point, and the angle is sustained by the two tangents. This surveyed angle, however, might be distorted on the map. For instance, the reconstruction of the surveyed angle between three points yields different values from the Gnomonic and the Mercator map projections (Figure 4.5), and the one from the Gnomonic projection is wrong.

4.5 Preserving Some of the Geometric Properties

It is not possible to completely preserve the geometry in a map projection. It is wishful thinking that if we cannot preserve length, angle, and area at the same time, we can perhaps preserve at least one of them. Unfortunately, this is not possible for length.

Preserving length means that the length measured along any line (curved or straight) on the map, after applying the scale factor of the map, should yield its true length on the ground. If a map could be developed to behave like this, which is impossible, it would have been called a map of constant scale. In reality, maps are enlarged in certain directions and reduced in others, with only lines observing some strict conditions being true to scale. Those map projections that preserve correct scale along great circles from one or at most two points are called *equidistant* map projections.

Preserving angles means that surveyed angles on the ground can be reconstructed correctly on a map using the method described in section 4.4. Map projections with such characteristic are called *conformal* (or *orthomor-*

phic). For this reason, conformal maps are particularly popular for topographic and engineering maps.

Preserving area means that the area measured on the map, after applying the scale factor of the map, should yield its true area on the ground. Fortunately, it is possible to develop map projections that preserve area (meaning that they show a constant scale in area) over the entire map. Such a map projection is called *equal-area* (or *equivalent*).

A map projection can be equidistant, conformal, or equal-area, understanding that the term equidistant must be interpreted in a restrictive sense. It is not possible, however, for a map to preserve more than one of these properties. For instance, no map can be equidistant and equivalent at the same time. Some projections, such as the Gnomonic, do not preserve any of length, angle, or area. That is why the distance measured along the straight lines representing great circle routes on a Gnomonic projection is not necessarily correct.

It would appear that for any map projection, geometric distortions would become increasingly smaller as the mapped area is reduced in size and would eventually become distortion free when the area becomes infinitely small. The surprising thing is that even this is not guaranteed. For instance, if a map is not conformal, angular distortion will not approach zero as the area gets smaller. Under certain circumstances, some of the distortions will approach zero as the area becomes infinitely small. We will touch on this again later.

4.6 Minimizing Geometric Distortions for a Project Area

It is now clear that it is impossible to eliminate geometric distortions on a map; the best we can hope for is to minimise them. Distortion minimisation is based on one simple principle: the place where the mapping plane (the flat paper) intersects the reference body (sphere or ellipsoid) is distorted the least. The amount of distortion will increase away from the place of intersection. The choice of a map projection for a certain region is concerned with generating an area of controlled distortion enclosing that region.

Take the case of a planar mapping surface. Its contact with the reference body is a point (called a *standard point*). As explained in section 4.3 and illustrated in Figure 4.1, distortion increases radially from the standard point. Hence, we can plot isolines of constant distortion as concentric circles (Figure 4.6) regardless of the location of the standard point and the shape of the graticule. The amount of distortion in a band between any two isolines will not exceed the higher of the two values. If the area to be mapped is shaped like a circular arc, then it could fit snugly between two isolines, thereby ensuring that the area is within an optimal region with distortion not exceeding a certain maximum. A mapping surface touching the reference body is called a *tangent* surface.

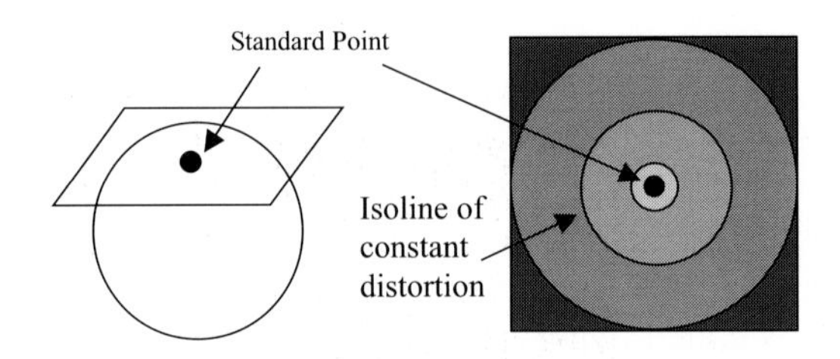

Fig. 4.6 Distribution of distortion for azimuthal projections

It should be noted that within the infinitely small area at the standard point, distortions could still occur depending on the map projection. If the map projection is not conformal, angular distortions will become extremely small but will never disappear.

As mentioned in section 4.3, map projections based on a planar surface have a special characteristic, which is that directions (or *azimuths*) from the projection centre to any other point on the map are correct. This class of map projections is therefore called *azimuthal.*

The point of contact can be enlarged to a circle of contact if the mapping plane cuts into the reference sphere, giving rise to the *secant* case. In this case, places with least distortion will be extended from a point to a line (a *standard line* or *standard parallel* if it follows a parallel). Isolines of constant distortion will still be concentric bands, but a band of the same size as in the tangent case will have a smaller distortion (Figure 4.7). The net effect is that the maximum distortion within the mapping area will be reduced compared to the tangent case.

Again, the standard line is not entirely distortion free. For any map projection, the scale along the standard line is true. The amount of the other distortions depend on the map projection.

On azimuthal maps, the projection centre at the centre of the earth is projected to the point at which meridians converge whether the aspect is tangent or secant. That point, which is usually at the centre of the map, is therefore also called the *projection centre.*

For areas that are not circular in shape, planar mapping surfaces will not be suitable. There are other types of mapping surfaces (behaving like planar surfaces) that intersect the reference body in different ways. They are called

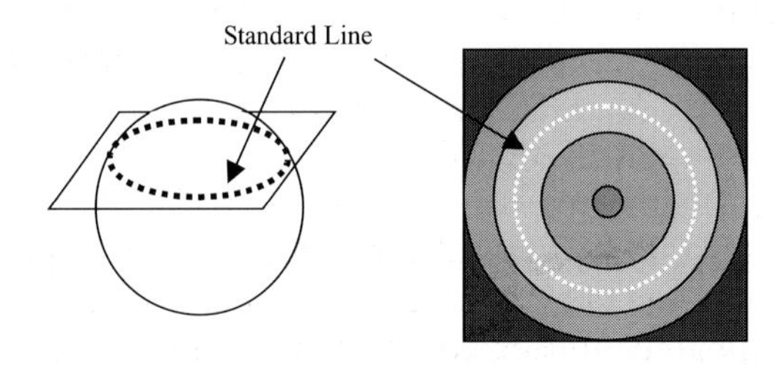

Fig. 4.7 Distribution of distortion for secant azimuthal projections

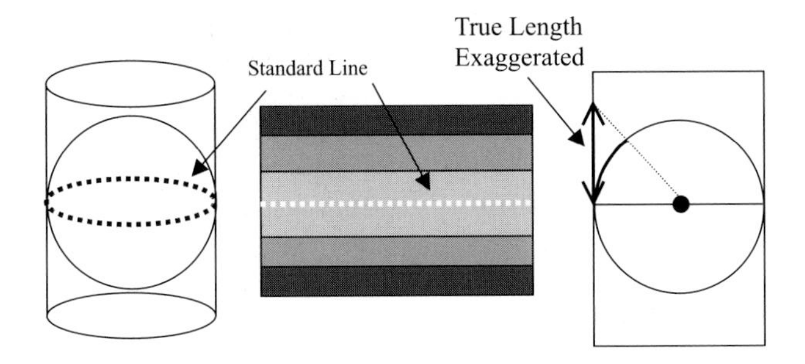

Fig. 4.8 Distribution of distortion for cylindrical projections

developable surfaces meaning that they can be developed (unfolded) into a flat surface without any geometric distortion.

Two developable surfaces, the cylinder and the cone, are the most popular for map projections. A cylinder tangent to a reference body will touch it along a great circle. The isolines of constant distortion will be bands parallel to the standard line (Figure 4.8). Hence, such bands are optimal for regions that are elongated and parallel to great circle lines. Secant cylinders cutting into the reference body create two standard lines (no longer great circles) parallel to each other. Similar to the case of a secant plane, a secant cylinder reduces the magnitude of distortion within a band of the same size. Map projections that are based on cylinders are called *cylindrical*.

A *conical* projection uses a cone as the mapping surface. A cone tangent to a reference body will touch it along a circular line other than a great circle. Isolines of constant distortion, like the cylindrical case, are concentric bands (Figure 4.9). They are optimal for elongated regions not following great circles. Again, secant cones generate two standard lines that tend to reduce the distortion within a region.

The plane, the cylinder, and the cone give rise to the azimuth, cylindrical, and conical map projections respectively. It is useful to note that a cylinder can be obtained by stretching the pointed tip up towards infinity, and the plane can be obtained by pressing that tip down to meet the base. The mapping surface can either touch the reference body (tangent case) or cut into it (secant case). The type of mapping surface and the way it intersects the reference body can be used together to create an optimal zone of acceptable distortion covering regions of different shapes and orientations. The map projection so produced could be equidistant, equal-area, conformal, or none of the above. Since no maps are of constant scale, the scale given for a map is applicable only at an infinitely small region at the standard point or along the standard line.

The last parameter we need is where on the reference surface the intersection happens. This is called the *aspect* of the mapping surface and there are three types: normal, transverse, and oblique. In the *normal* aspect, the axis of the mapping surface coincides with the north-south axis of the earth. This aspect will generate the simplest graticule and will simplify

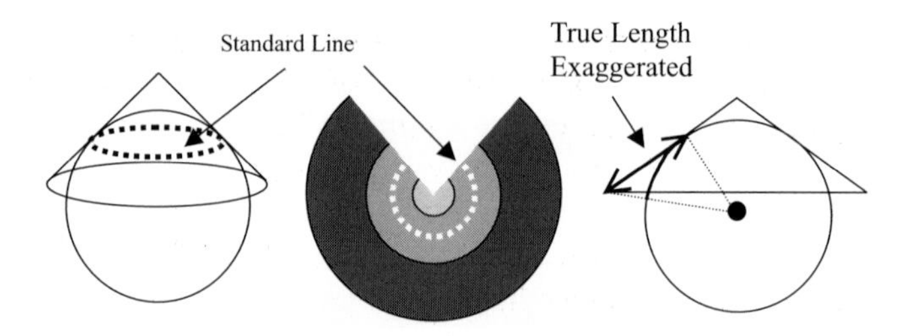

Fig. 4.9 Distribution of distortion for conic projections

the calculations. In the normal azimuthal case, the plane touches one of the poles and is also called the *polar* aspect. The meridians in this case are straight lines and the parallels concentrate circles, assuming that the reference body is a sphere. In the *transverse* aspect, the axis of the mapping surface is orthogonal to that of the earth. For azimuthal projections, this is sometimes called the *equatorial* aspect. In the *oblique* aspect, the place of contact is neither the pole nor the equator.

4.7 Naming of Map Projections

The four major properties of map projections; namely, mapping surfaces, intersections, aspects, and geometric properties, affect the characteristics and applications of those projections. To describe a map projection, we might have to use something like "conical tangent normal equal-area." In practice, we normally skip the part on intersection and aspect if they are non-consequential or if they are "tangent" and "normal" respectively. Hence, the above could be abbreviated to "conical equal-area." When these properties are not enough, we often add the name of the inventor. For instance, the "Albers equal-area conical projection" is different from the "Lambert equal-area conical projection" in that the first maps the North Pole as an arc of a circle, while the second maps it as a point. For some special cases, the inventor of the projection is sufficient to uniquely identify it. The most famous example is the Mercator projection, which is conformal and cylindrical. Incidentally, the Mercator projection is also called the Wright projection.

4.8 Rectangular Grid System

The geographic graticule on a map is not convenient for making measurements because it is not necessarily comprised of straight lines and the units are in the less convenient degrees of latitude and longitude. For topographic maps, it is useful to superimpose on the map a rectangular grid in ground units of say, metres. The main purpose of the grid is to allow linear interpolations to read ground coordinates of points from which angle, distance, and area can be computed. When calculating azimuth where the direction of north is important, it is necessary to distinguish between the *grid north* from the rectangular grid and the *true north* from the meridians.

For any coordinate system we need an origin. For convenience, the origin of a rectangular grid system superimposed on the geographic graticule is located at where the *central meridian* meets the standard parallel at the centre of the region being mapped. When there are two standard parallels (as in secant cylindrical and secant conical projections), another parallel in between will be used called the *reference parallel.*

A normal origin will have $(0, 0)$ as its coordinates. This, however, will generate negative coordinates in some parts of the region. Negative coordinates are not desirable because the dropping of the sign creates serious errors in calculations. To avoid this, the origin will be given a large positive number. This effectively shifts the origin towards the lower left corner of the region. The original origin in this case is called the *false origin* and its coordinates *false coordinates.*

4.9 The UTM Projection

If topographic maps from neighbouring countries have different map projections and rectangular grid systems, measurements across the border would be difficult (Figure 4.10). This problem was realised particularly during World War II when cross-border warfare was common.

After the War, the nations of the North Atlantic Treaty Organisation (NATO) decided to adopt a common coordinate reference system and a common projection for mapping at topographic scales. The system chosen was the Universal Transverse Mercator (UTM) Projection for most parts of the world (a different map projection was chosen for the polar regions). It is a method of systematically mapping the entire world using a Transverse

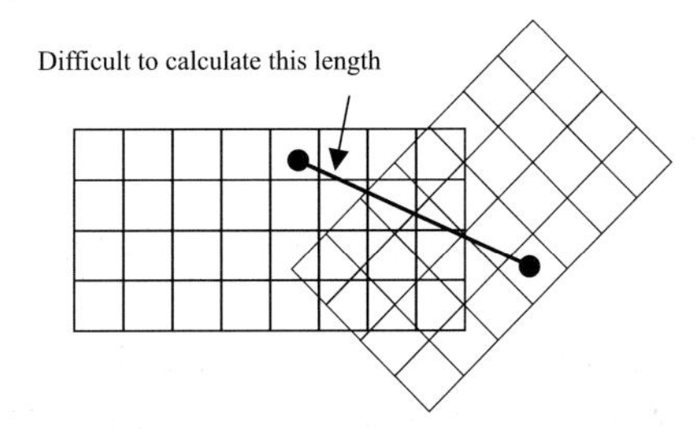

Fig. 4.10 Confusion caused by mismatching grids

Mercator projection, and the UTM was chosen to meet the following major design specifications:

a) The projection should be conformal.
b) Scale errors are not to exceed a specified tolerance suitable for large to medium scale applications.
c) Angular errors are not to exceed 5 degrees.

It is impossible to map the entire world onto a single surface satisfying the above requirements, but mapping a smaller part of it could be done. The scheme therefore is to break the world into parts (or *zones*) and use a UTM projection to map each zone (Figure 4.11). This is done by rotating a cylinder around the earth at fixed intervals, and mapping a zone at every interval. It is true that this would not produce continuous coverage of the entire world in a single coordinate system, but at least the transition from zone to zone is systematic, thereby simplifying the mathematics involved in converting coordinates from one zone to another.

In theory, a normal Mercator or a transverse Mercator projection can be used. The transverse cylinder offers an important advantage over a normal one – it is in contact with the two poles and meets the reference body along a meridian. A zone, therefore, can be delineated conveniently surrounding a meridian from pole to pole. Using a normal cylinder, the poles will be projected to infinity. Rotating the cylinder in the transverse case will cause it to come in contact with different meridians, eventually covering the entire world with a number of non-overlapping zones. The design specifications of the UTM projection dictate that the zones are 6 degrees wide in longitude, thus dividing the world into 60 zones. This requires that the cylinder, turned 90 degrees relative to the north-south axis, be rotated around the world 6 degrees at a time.

This projection preserves all the characteristics of a Mercator projection except that a loxodrome is no longer a straight line. The graticule within each zone no longer forms rectangular grids: the parallels are ellipses, while

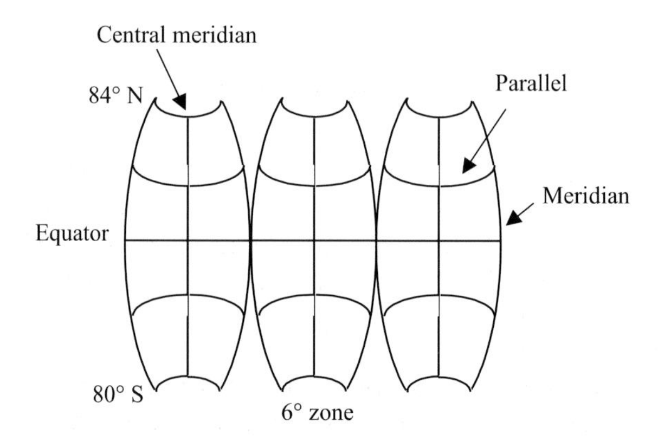

Fig. 4.11 UTM Zones

the meridians are complex curves. Nevertheless, the graticules in every zone are identical in shape, which is mathematically desirable and cartographically convenient. One template of the graticule can be used for all zones.

If the cylinder were tangent to the reference body, it would be in contact along a meridian at every zone. This is called the *central meridian* because it is at the centre of the zone. In reality, the cylinder cuts into the zone to further reduce the scale error. As a result, there are two standard lines within each zone on either side of the central meridian. Consequently, the scale along the central meridian will not be true, but reduced. This is expressed as a *scale factor* for the central meridian and for the UTM projection it is 0.9996. Adjusting this factor effectively changes the amount that the mapping plane cuts into the reference body. This factor increases to about 1.00095 at the edge of the zone. In other words, scale distortion within a UTM zone is no less than 0.0006. An approximate equation for computing the scale factor is

$$s = s_0(1 + X^2/2R^2),$$

where s_0 is the scale along the central meridian,
X is the ground distance from the central meridian, and
R is the radius of the earth.

The zones do not actually include the poles. They extend from 80°S to 84°N, leaving the polar regions to use the Universal Polar Stereographic (UPS) Projection, an azimuthal projection with the projection centre at the anti-pole, a point 180° from the pole. The main reason for not including the two poles in an UTM zone is that this would cause the two polar regions to be partitioned into 60 zones, which is too many to be practical. Using a UPS projection, each polar region is covered by two zones each in the shape of a half circle.

Modified versions of the UTM have been used for regional mapping. A three-degree UTM using three-degree instead of six-degree zones would be useful for smaller regions by providing lower degree of distortions.

4.10 The UTM Grid

The meridians and parallels on a map in the Transverse Mercator projection do not form rectangular grids. It is therefore useful to create a rectangular grid system on top of it just for making measurements (Figure 4.12) as explained in section 4.8. When designing the UTM projection, a corresponding grid system was also developed. Again, this ensures a uniform grid system from zone to zone.

Each zone is covered by one grid system. Within a zone, the only straight lines in the graticule system are the equator and the central meridian, and they form the natural axes of the rectangular grid system (Figure 4.12). At

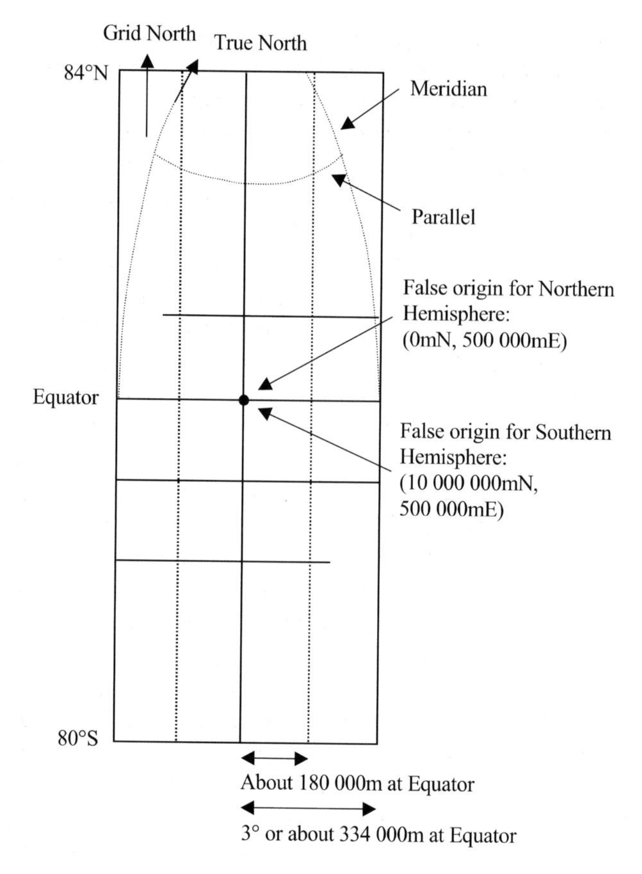

Fig. 4.12 UTM grid system within a zone

the four corners of the zone, the maximum deviation of the grid lines from the graticule occurs. As stipulated in the design specification of the UTM projection, this deviation (called *convergence of meridian*) should not exceed five degrees.

Note that the reason for creating a grid system is to help users manually read coordinates quickly with the least possibility of making errors. This is particularly important for military operations where locations must be read and transmitted quickly and correctly. The X and Y coordinates are abstract concepts and they are replaced by metres in the north direction (metres northing or mN) and metres in the east direction (metres easting or mE).

Long coordinates are more error-prone than short ones simply because there are more digits involved. The size of each coordinate can be reduced if we confine the measurements to a smaller portion of the entire zone. So instead of using one origin for the zone and requiring long coordinates to reach its four corners, we can break the zone into cells and have a local origin for each. If the smaller cells are only 100 000 m long on both sides, then the largest coordinate required to reference points in each cell is only 99 999 m at a precision of one metre. Note that the coordinate 100 000 m is not required because it is just 0 m of the next neighbouring cell. One problem is that these local coordinates are unique only within a cell and could be repeated in other cells in the same zone and in other zones. As long as the calculations do not cross a cell, there is no need to identify each local coordinate with the zone in which it falls.

The UTM grid system employs a rather extensive scheme to subdivide the entire zone into progressively smaller cells and to name each cell in a unique way. The smallest is a matrix of cells 100 000 by 100 000 metres large each of which is identified by a two-letter code, such as CA (Figure 4.13). The first letter is a column designation, while the second is a row designation. These codes are not unique and will be repeated about every nine degrees of longitude and latitude. To make such squares unique, their code must be preceded by an identification of the higher-order cells in the zone containing these smaller cells. The higher-order codes are called *grid zone designations*, such as 1N. The first number in this designation represents the zone, which is numbered starting from 1 for the zone between 180°W and 174°W of longitude and increases towards the east. The second letter is for a belt that covers 8° of latitude (Figure 4.13) except the north most one, which covers 12°. There are a total of 20 such belts covering the area from 80°S to 84°N, and these belts are labelled from C to X skipping I and O for fear of confusion with zero and one. Zones A and B are reserved for the South Pole region, which is covered by the UPS. Zones Y and Z are reserved for the North Pole region also covered by the UPS. To give unique coordinates to a point world-wide, the full designation would be something like 1N CA 08 375 mN 74 562 mE accurate to one metre on the ground.

For calculations involving points that are far apart, a zone-wide coordinate system without the use of the grid zone designation and the 10 000 m

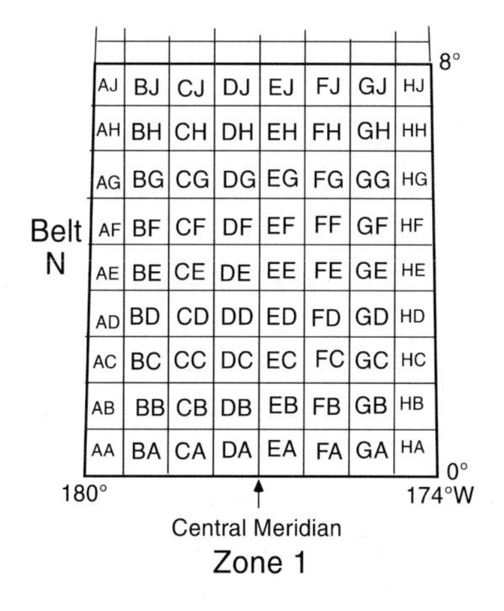

Fig. 4.13 Grid zone designation and the 100 000 m squares

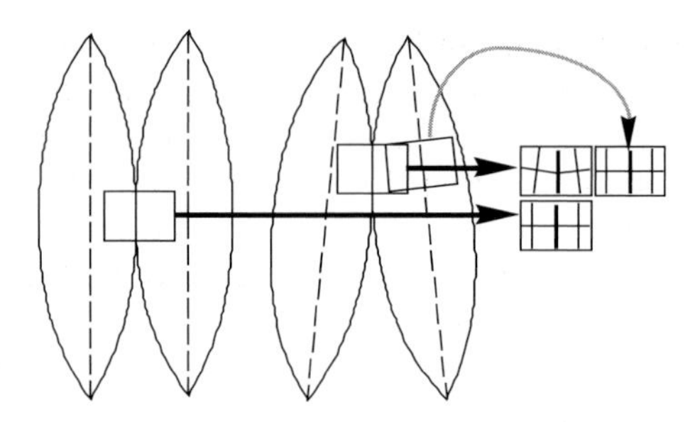

Fig. 4.14 Maps crossing UTM zones

square designation is desirable. An ideal location for the origin is the centre of the zone where the equator meets the central meridian. As explained in section 4.8, this false origin is assigned with *false northings* and *false eastings* to avoid negative values.

In the UTM grid system, the false origin serves two hemispheres (Figure 4.13). For the Northern Hemisphere, its value is (0 mN, 500 000 mE) since the origin has been shifted to the lower left corner of the Northern Hemisphere. For the Southern Hemisphere, its value is (10 000 000 mN, 500 000 mE) for the same reason. The breaking of the zone along the equator effectively reduces the size of the coordinates in each hemisphere. For calculations in a zone across the hemispheres, it is easy to convert Northern Hemisphere coordinates to agree with those in the Southern Hemisphere by adding 10 000 000 m to northings.

The grid system is unique within a zone. When two zones meet on a map, two grid patterns will meet at an angle (Figure 4.14) making measurements across a zone difficult. To get around this problem, either of the grid systems can be extended 40 km to one side. Geometric distortion increases beyond the UTM specification within the extended zone, with the scale factor increasing to about 1.001 at the edge.

4.11 Integrating Maps of Different Projections

When creating a geographic database, it is often necessary to combine maps in different projections. Normally, the database is already in one map projection and the new data in another. A simple and easy (but not very accurate) method is to assume that the earth is flat over the project area. If the area concerned is small, the curvature of the reference body could be small enough to be ignored. In this case, we can register the new map over the one already in the database using one of the simpler transformation methods, such as affine or projective, described in Chapter 3 of this book.

This method is not accurate because we have assumed that the earth is flat. The proper method is to perform a map projection conversion mathematically, and this can be done by most mapping software.

A map projection has a forward set of equations that will transform from the geographic to the Cartesian system: $f(\phi, \lambda) \rightarrow (X, Y)$. It also has an inverse set of equations that will convert from the Cartesian to the geographic system: $f^{-1}(X, Y) \rightarrow (\phi, \lambda)$. It should be noted that although the measurements based on Cartesian coordinates on a map are not accurate due to map projection distortions, these coordinates can be converted back to accurate latitudes and longitudes.

The conversion from map projection f to g can be thought of as two separate transformations:

$$f^{-1}(X, Y) \rightarrow (\phi, \lambda) \text{ followed by}$$
$$g(\phi, \lambda) \rightarrow (X, Y).$$

The projection functions f and g (and their inverses) require that the user supply values to a number of variables (called *map projection parameters*), such as scale and location of the standard point or line. The parameters required depend on the projection. The mandatory parameter is the reference body usually identified by its name such as Clark 1866 or WGS 1980. Information that is dependent on the projection includes the location of the standard point, standard parallel(s), central meridian, and scale factor at the standard point or parallel line. Most of the projections would need the false origin (false northing and false easting) in order to generate the rectangular grids. In that case, the longitude (central meridian) and latitude (either standard parallel or reference parallel) of that origin will also be required. If these parameters are not shown on the document, then the responsible mapping authority should be consulted.

The following shows the parameters of a number of common map projections.

UTM

a) Location of the central meridian (or the zone number).
b) Scale factor of the central meridian to determine how far the cylinder cuts into the reference body. This is 0.9996 but could be different for modified versions of the UTM.

Albers Equal-Area Conical (normal aspect)

a) One or two standard parallels.

Stereographic (Azimuthal)

a) Latitude and longitude of the projection centre.
b) Scale factor at the projection centre to determine how far the plane cuts into the reference body. This is 1.0 for the tangent case and less than 1.0 for the secant case.

4.12 Map Projection and Digital Maps

Traditionally, error analysis for map projections has been a major concern. This is because of the important role maps have played as analytical tools. With the growing popularity of digital maps, the need for a map projection to minimise geometric accuracy has been reduced. Calculations requiring high precision need no longer be performed on a flat map surface but can be done analytically using stored coordinates. Even if the stored coordinates are based on the Cartesian system, they can always be converted back to accurate latitudes and longitudes for computations to be done on the ellipsoid. The computer can perform these conversions very fast in real time.

There is still a need for map projections even when the digital maps are stored in the geographic coordinate system of latitudes and longitudes because a flat surface is still used for display. A very simple method, which is a trivial projection, is to plot the geographic coordinates as Cartesian coordinates. In other words, the point longitude = 102°30′E, latitude = 42°30′N will be plotted at X = 102.5 units and Y = 42.5 units. For systems that store points in geographic coordinates, this is usually the default projection for display. The user would need to identify a projection and its parameters if another map projection is desired. In fact, when the scale of the map is large, any map projection will do.

When the scale of the map is small, particularly when a global map is needed for thematic mapping purposes, the cognitive impact of a map projection has to be considered carefully. For instance, it would be visually misleading to use a conformal map projection for dot maps where the density of the dots relative to the area of the regions is important. An equal-area map projection would be a better choice.

The UTM creates a unique problem for some digital mapping software. This projection does not employ a single coordinate system but uses a distinct system within each of the 60 zones although the transformation between zones is uniform. Hence it is rather difficult to digitise a map that covers two UTM zones. The two zones involve two central meridians and two origins. To handle this problem, each point must be identified with a zone number and its hemisphere so that its correct position can be determined. This is an additional burden on storage and digital mapping software not all provide this flexibility.

Even if a system can handle points from multiple zones, there is still a problem of gaps between zones. We normally do not see gaps on a map covering two zones because the zones are rotated to meet before the map is made (Figure 4.14). Within a map, the curvature of the zone boundary is too small to be noticeable. The gap will become obvious when maps along the boundary are laid out on a (very large) table. This is because although the zones cover the entire earth, together they do not completely cover a rectangular area. A truly universal coordinate system is based on latitudes and longitudes.

4.13 Conclusions

We have shown that although flat maps are never geometrically 100 percent accurate, they are useful display and visual tools. Different kinds of map projections have been and will be developed to serve various applications. We have reviewed the various parameters affecting the characteristics of a map projection, and have described the UTM system in more detail.

The UTM is the most popular projection for topographic mapping. It was developed to provide coverage of the world in a systematic series of map projections and rectangular grid systems. We have shown that this system, although good for manual analysis, could cause problems for some of the mapping software when digitising areas that cross UTM zones.

We have also seen the need to convert maps from one projection to another, particularly when integrating data from different sources. In order to do this, we need to know the exact parameters of each of the map projections involved.

5 Geographical Data from Analogue Maps

Yuk-Cheung Lee and Lilian Pun

5.1 Introduction

Analogue maps have been a popular source of data for GIS. When an organisation first develops a GIS database, analogue maps are often the main source of data because digitising them is usually more cost effective than collecting data using land surveying, photogrammetry, or other data acquisition methods. Although data digitised from maps are usually less accurate, we can skip many of the elaborate and tedious steps of undertaking measurements on the ground and on stereo models. Moreover, collecting data from a map can cover much more ground in much less time, without any accessibility problems.

In this chapter, we will discuss the acquisition of geographical data from analogue maps. By geographical, we mean both positional and thematic data representing, respectively, the geometric and non-geometric characteristics of geographic features. Many of the concepts also apply when the document to be digitized is not a line map but a photographic image, such as an aerial photograph. For special characteristics of aerial photographs, please refer to Chapter 9.

5.2 Vector and Raster Models

In conventional mapping, we often compare the use of line maps and orthophotographs. An orthophotograph is obtained from an aerial photograph of an area by correcting its geometric distortions due to terrain, the instability of the airplane taking the photograph, and other factors as explained in Chapter 10.

A line map is the result of interpretation meaning that features have been identified and delineated, mainly by lines, on a map. Line maps are rather sparse documents with more empty space than lines. An orthophotograph, being fundamentally a photographic image, has not been interpreted. Users must therefore use their knowledge and experience to recognize features on an orthophotograph. Compared to a sparse line map, an orthophotograph has no empty spaces because each spot on it has a colour or gray level.

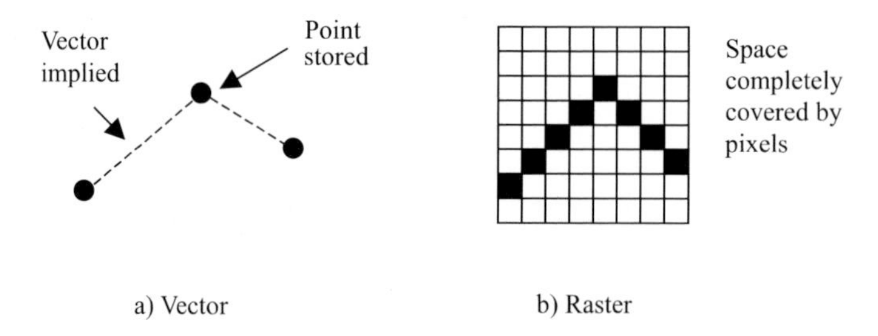

Fig. 5.1 Vector and raster representation

The same comparison holds for vector and raster maps, which can be regarded loosely as digital equivalents of line maps and orthophotographs. In the *vector* model, the outlines of features are represented by a series of points, which are connected by straight-line segments called vectors. These vectors are not actually stored in the database except for the end points (Figure 5.1a). We see the vectors only when the points are joined during display, and the vectors are assumed in the calculations, such as finding the length of a line.

With the *raster* model, the mapped area is completely covered by a matrix of identical (usually square) cells called *pixels* (Figure 5.1b). Each pixel is associated with a code that could have a wide range of meanings. If the raster data were obtained from photographing the terrain, then each pixel would store the reflectance level of the terrain assigned by the sensing device (Chapter 9). For a monochromatic system, this will yield a range of gray levels. For a colour system, this will yield a range of colour values. On the other hand, each pixel could be associated with a code that reflects the thematic nature of the geographic feature upon which it falls. For instance, we can label a pixel as "river" to indicate that it falls on a river. There are methods of converting reflectance levels to a corresponding thematic value almost automatically, and they are discussed in Chapter 12.

In this chapter, we are particularly interested in converting a line or choropleth map on paper into either a vector or raster representation in the computer. If we convert a line map into vector format, then a continuous line on the map becomes a series of points. Attributes of the line would be associated with its list of points. If we convert a line map into raster format, a collection of pixels will result, and each pixel would be associated with a code to indicate the colour of the map at that spot.

In both vector and raster models, the concept of resolution applies. *Resolution* of a database is the smallest detail we can represent in the database. In the vector model, it is determined by the precision of the coordinates (Figure 5.2a). Naturally, the precision of the coordinates could degrade the accuracy of the data if the resolution is lower than its original accuracy. A vector database that can store coordinates up to a millimetre is higher in resolution than one that can store only up to a centimetre because database with centimetre resolution cannot represent a line accurately that is

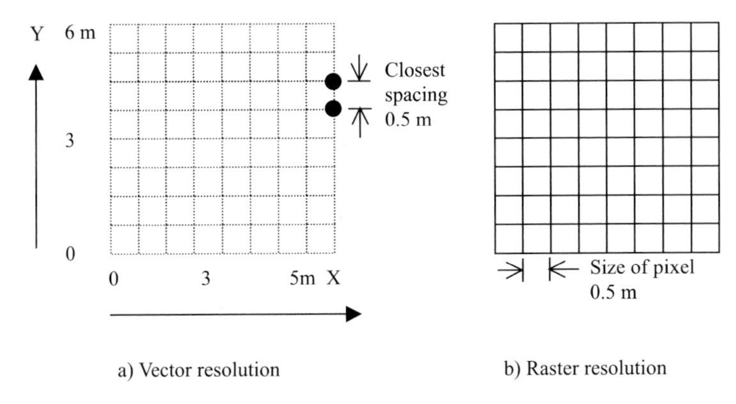

Fig. 5.2 Vector and raster resolution

shorter than a centimetre. In a raster model, resolution is determined by the size of the pixels (Figure 5.2b). A database with a pixel size of one millimetre square is higher than one with a pixel size of one centimetre square. In the latter case, an area smaller than one centimetre square cannot be represented accurately.

5.3 The Choice between Vector and Raster Models

The choice of converting a paper line map into a vector or raster form is affected by the following factors.

Data Storage

A geographic database in the vector model is often smaller in storage size than that in a raster model. One reason is because a vector map is sparser than a raster map for which every location must be represented by a pixel. This, however, cannot be considered independent of the database resolution.

Resolution

If a vector and a raster map were of the same resolution, then the raster version usually would be larger in volume. Moreover, the capacity of a raster database increases to four times if its resolution doubles. For practical reasons, the resolution of a raster database is often compromised in order to keep storage to a manageable size.

On the other hand, the effect of resolution on the size of vector maps would be less. Using 32-bit integers to store each coordinate, we can cover the entire world with a resolution of about one centimetre. When the area to be covered is much smaller, the maximum resolution can be increased. In other words, 32-bit integer coordinates are precise enough for most applications, and hence the resolution of a database could be varied in a very wide range without any increase in storage size.

Speed of Data Capture

A vector map is usually produced by manually tracing a line map, while a raster map could be generated automatically by means of a scanning device. There are hybrid methods that could produce a vector map semi-automatically. In general, it takes longer to produce a vector map than a raster map.

Application Software

Data captured from a map is ultimately used in GIS software. Each piece of software assumes either the vector or the raster model as its prime model for processing and analysis. The target application software then often dictates the model required.

The detailed reasons why GIS software chooses either the vector or the raster as its prime model are beyond the scope of this book. The essence of them, however, can be related back to the differences between a line map and an orthophotograph. Vector-based GIS software prefers the use of well-defined and delineated geographic features because it makes certain operations easier. One of them is the calculation of topological relationships (such as one feature being surrounded by another) among geographic features. The raster-based GIS software prefers the entire database to be completely covered by pixels (or data) because it can identify very easily the thematic nature of a given location. To do this, the software needs to identify only the pixel the given location falls on, and then retrieve the stored value for that pixel. In a vector case, the given location often falls onto the "empty" space between points, and more involved geometrical calculations must be performed to answer the same question.

5.4 Vector Digitizing

The basic unit in a vector format database is a point. Lines are composed of an ordered sequence of points, and lines delineate the boundaries of areas. One unique advantage of digitizing geographic data from maps is that there is very little need for interpretation. Features are already clearly defined, and some of their attributes are shown either by symbology or by annotations. Selection might be required if not all the features from a map are useful, and additional attributes from other sources might be included. Unless the map requires updating, there is no need to add or modify the features. Hence the task of digitization in the vector format mainly involves tracing the feature outlines and keying in their attributes. Due to the amount of manual labour involved, this could be a time consuming process.

Digitizing a map is to convert its cartographic elements into a form compatible with computer processing. The cartographic elements are symbolized points, symbolized lines, symbolized areas, and annotations. As we have mentioned before, symbology carries attribute information about the features, a consequence of the dual purpose of maps serving as data storage and as a data display medium. Before we discuss the process of digitizing, we will first review how symbology could be handled in a GIS.

5.4.1 Representation of Symbology in Vector Format

In a digital database, the position of a feature is often separated from its cartographic representations. In other words, the geometry of features is stored in unsymbolized form while their attributes are stored as alphanumeric data. Hence, in digitizing a map, we ignore symbology while tracing the outlines of features, and enter their attributes separately. For example, a dashed line denoting a footpath will be traced as a continuous line, while the fact that it is a highway will be coded in a form stipulated in the design specification of the database. In particular, we need to know the code that means a "footpath" in the database. The design of database organization and the standard for all the necessary codes is an important task before one can start populating the database with data. Two popular attributes are feature type code (also called *feature code*) and feature identification code (also called *feature key*). The feature code identifies the class to which the feature belongs, whereas the feature key uniquely identifies a feature.

During the display of the digital map, the symbology will be reconstructed from its attributes. For example, a footpath will be shown as a red dashed line (Figure 5.3). The display software performs this by referring to the attributes of the features and consulting files that control how the symbology is to be applied. These files for symbolization rules are often created for each series of maps indicating how every type of point, line, or area feature is to be symbolized. These rules serve a similar but reverse function of legend boxes

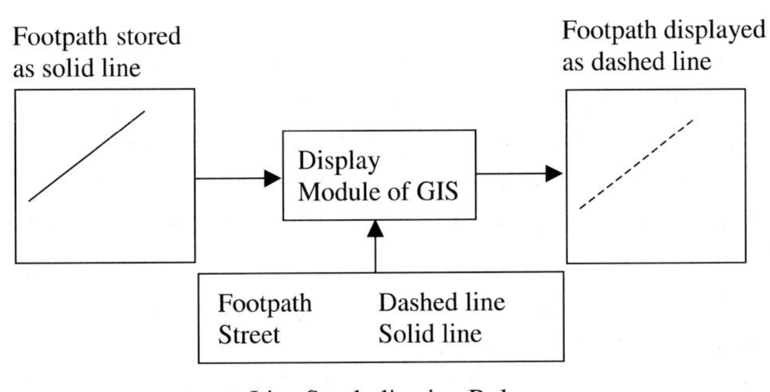

Fig. 5.3 Symbolization of geographic line features

on conventional maps. They indicate the symbol to be applied to each type of feature instead of explaining the meaning of each symbol.

5.4.2 Digitizing Point and Line Features

To digitize a point feature, it is only necessary to record its location, usually in two dimensions. Attributes about the feature, such as its feature code and identification, will also be entered manually. It is not common to capture three-dimensional data from flat maps although it is possible to interpret height from contours. Some point features, such as spot heights above water and soundings below water, do contain data in the third dimension. The elevation and depth will be entered by hand.

To convert a graphic line (straight or curved) into digital form, it is necessary to break it up into a series of discrete points. The closer the points, the better the approximation to the original curve. An optimal spacing would be one that would generate vectors that are not noticeable as straight segments at the display scale. These vectors can be longer on smooth curves than on tight ones, but normally not much shorter than 0.1 mm.

A practical problem in applying this rule is deciding on the scale upon which one should base the optimal spacing of points since a digital map could be displayed at a wide range of scales. A paper map, the document upon which digitizing is performed, has a scale, which in turn is associated with an indication of accuracy. When we enlarge a map, we effectively exaggerate its positional error. It would not be a good practice to display a map at excessively large scales, and the spacing of points can therefore be computed from the maximum output scale appropriate for the accuracy of the input document. When the database is displayed at a much smaller scale, the points will become too close to each other or even overlap. In that case, the redundant points need to be removed through a filtering process.

5.4.3 The Digitizing Table

A digitizing table is a device used to manually convert graphic lines into coordinates in digital form. It consists of a fine grid of wires embedded between flat glass or plastic plates. The wire mesh determines the *table coordinate system*, which is one of two-dimensional Cartesian. The unit of measurement is traditionally in inches or millimetres, and the resolution is normally 0.001 inch (0.025 mm) for tables used for mapping purposes. The user can determine the origin of the system, and it is usually set at the lower left corner of the manuscript so that only positive coordinates are generated. On top of the surface is a free-moving tracking *cursor*. On the cursor is a pair of crosshairs to indicate the point at which coordinates are read, and around the cursor is a magnetic coil that senses the wire mesh to compute position.

A keypad, usually with 16 buttons, is also included to allow input of values and commands. The cursor is connected by a cable to a controller, which is responsible for converting signals from the coil to the table coordinates, to handle communications with the computer, and to control the operation of the digitizer.

Digitizing in vector format can be done in either point or stream mode. In *point mode*, a single point is passed to the host computer at the press of a button on the digitizing cursor. It is useful for digitizing point features and points on angular lines, such as building outlines. Using point mode to digitize a curved line is a very tedious process because points on the line must be digitized one by one, and there could be a large number of them.

In stream mode digitizing, the digitizer hardware invokes an automatic sampling process and generates points automatically. Sampling can be according to distance or time, but time sampling is much more common for digitizing maps. Using stream mode, the operator needs to press the digitizing button only to start and end a line. The points in between will be generated automatically while the operator tracks along the line. Using time sampling, the spacing of the points will be regulated by the tracking speed. This is desirable because operators normally track faster on smooth curves generating longer vectors and slower on tight curves generating shorter vectors. The average tracking speed is about 1.5 mm per second. In order to generate points no wider than 0.1 mm apart, a sampling rate of about 15 points a second should be used. In practice, this rate is increased to about 30 points per second to produce even closer points so that the lines will still appear smooth at larger output scales.

5.4.4 Accuracy of Manual Digitizing

The accuracy of digitizing using a digitizing table depends on the accuracy of the original manuscript, pointing, and the device. The accuracy of maps is controlled by accuracy specifications developed and monitored by mapping agencies. High quality topographic maps of the world depict 90% of well-defined point features within 0.5 mm of their correct positions at map scale. The original manuscript, however, particularly one on unstable paper, might be distorted. Paper expansion and contraction due to humidity can span a range of +/– 0.05%. Over a length of 0.5 m, it can amount to 0.25 mm. These distortions could be corrected to some extent through the registration process as described in section 5.4.5.

The accuracy of pointing is affected by line thickness, cross-hair thickness, and our physiological limits. A thicker lines means that it is more difficult to position the cursor over its centre. The following are some typical line widths:

- Fine pencil line 0.13 mm
- Ball point pen 0.25 mm
- Fibre-tip pen 1.25 mm

The performance of the operator is highly dependent upon the skill, dedication, and stamina of the person. For example, the eye can resolve about 500 lines per inch (200 lines per centimetre) under bright illumination at about 10 inches (4 cm). Under very low-level illumination, this is about 100 lines per inch (40 lines per centimetre). To minimize this problem, some vendors offer backlighted digitizer surfaces. Accuracy is higher in point mode digitizing and lower in stream mode digitizing with an accuracy of about 0.2 mm.

Accuracy of the digitizing table also has an effect on the final result. The resolution of cartographic digitizing tables is usually 0.025 mm. With other mechanical imperfections, the expected accuracy is about five times the resolution, or 0.125 mm. This is sufficient for topographic maps considering the tracking accuracy and the thickness of the lines on a map.

5.4.5 Map Registration

When we digitize a point on a table digitizer, we measure the coordinates of the point in the *table coordinate system* described in section 5.4.3. This system is not very useful for data analysis because it can only provide measurements in the unit of the table, such as the length of a line being 3.5 cm on the table. This is just like reading coordinates and making measurements using a graph paper superimposed on the map. For the coordinates to be useful, they must be in the same coordinate system as that used by the map being digitized. That system is based on a map projection yielding units of usually metres on the ground and is called the *grid coordinate system*.

We need a set of mathematical equations to convert the points from the table system to the grid system. There are several classes of these equations representing different types of transformations to be explained later. Each class of equation has a set of parameters that must be solved before the equation can be used to convert points from one system to another. The process of solving for the parameters for a pre-defined set of equations is called *map registration*, a method of geo-referencing.

The steps involved in map registration are highlighted here (Figure 5.4). First we decide on the type of transformation. One of these types preserves the shape of the features after transformation and is called a *similarity* transformation. There are practical reasons why we sometimes do not want to preserve shape as explained in section 5.4.5.1.

This transformation has four parameters (say a, b, c, and d), and we will illustrate the process of registration using an example in Chapter 3. If we know that two points measured in the table coordinate system having ($x_1 = 102.571$, $y_1 = 44.461$, $x_2 = 99.403$, $y_2 = 242.910$) are converted to grid coordinates ($X_1 = 250$, $Y_1 = 200$, $X_2 = 120$, $Y_2 = 350$) respectively, we can

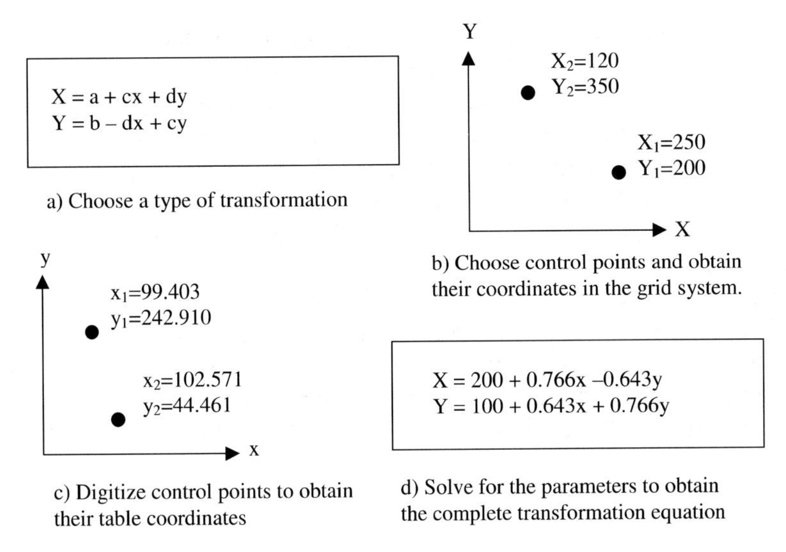

Fig. 5.4 The steps of map registration

determine mathematically the value of the four parameters ($a = 200$, $b = 100$, $c = 0.766$, and $d = -0.643$). These points with known table and grid coordinates are called *control points.*

After solving the parameters, the objective of map registration is fulfilled. Hereafter, every point digitized on the table reading (x, y) can be transformed using the same set of equations but with known parameters to yield (X, Y) in the grid system. For example, the point digitized as (110, 100) on the table will be transformed to (219.96, 247.33) in the database (Figure 5.5).

Map registration therefore relates two Cartesian coordinate systems, one for the table and one for the map projection used. We can imagine it to be the process of stretching one system on a rubber sheet to fit another, hence map registration is also known as *rubber sheeting.*

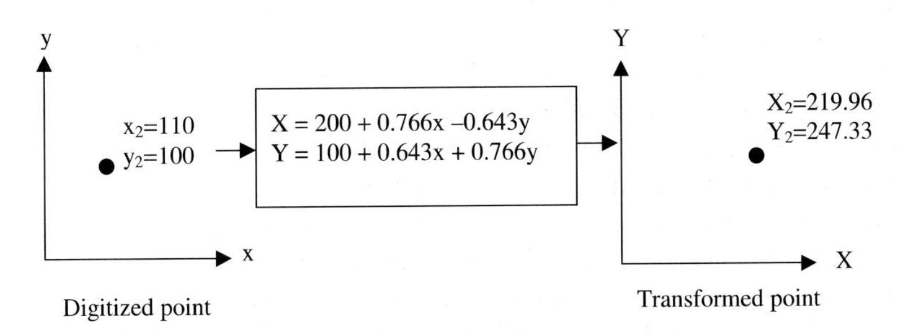

Fig. 5.5 Transformation of digitized point

5.4.5.1 Map Registration as Transformation Parameters Calculation

Mathematically, registration calculates the transformation between the table and the grid coordinate systems. The mathematical concepts of this have been explained in Chapter 3.

There are three transformations commonly used in table digitizing: similarity, affine, and projective, as mentioned in Chapter 3. Similarity preserves shape, affine preserves parallel lines, and projective preserves straight lines.

The main difference among them is the amount of geometrical distortion that can be corrected as a bonus of map registration. Since paper maps are not dimensionally stable, their neatlines are bound to be out of square. Transformation from the table system to the projection system can correct this distortion at the same time. A *similarity transformation* can correct a map that has been distorted uniformly so that its neatline remains as a rectangle perhaps rotated and scaled (Figure 5.6). An *affine transformation* can correct a map that has been skewed – the corners are out of square but the sides remain parallel. It includes the similarity transformation as a special case. A *projective transformation* can correct a map that has been stretched in different amounts from the four corners, but the sides remain straight. It contains the affine transformation as a special case, and is the most general of the three methods. These transformation methods would not

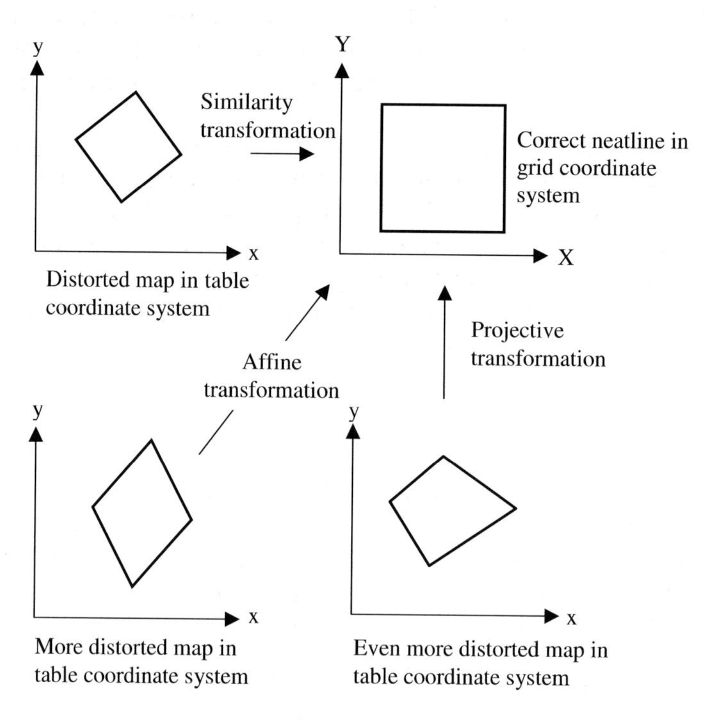

Fig. 5.6 Transformation types used for map registration

work if the straight lines have been distorted into curves. We will discuss registration for these maps in the next section.

The type of transformation to choose depends on the digitizing software used because some might not support all three methods. It depends on the area of the map to be digitized. In general, a very small area needs only the similarity transformation, while a large area needs the projective transformation. It also depends on the quality of the original map. Maps on more stable materials, such as film or scribe coat, would perhaps need only a similarity transformation.

The relationship between the table coordinate system and the map projection system is expressed as a set of equations with a number of *transformation parameters*. For instance, the four parameters of a similarity transformation are scale, shift in x direction, shift in y direction, and rotation. A registration operation computes the values of the four parameters through the help of two control points in two dimensions (four coordinates), which are known in both coordinate systems involved in the transformation. That is to say, the operator selects four points that can be identified on the map with known grid coordinates, either by interpolating from the map or from another source. Ideal controls are grid intersections from which coordinates can be read directly and triangulation stations for which accurate coordinates are available from the survey office. Other than that, any well-defined points, such as road intersections and house corners, can be used. To obtain the table coordinates for these points, the operator simply has to digitize them. After digitizing the two control points, the registration software will automatically compute the transformation parameters. An affine transformation needs three control points, while a projective transformation needs four.

Map registration involves the manual digitizing of control points and therefore can be affected by the accuracy of digitizing, as illustrated in Figure 5.7. In Figure 5.7, both control points are digitized incorrectly. Without knowing any better, the registration process assumes that they are correct and hence matches their wrong position in the table system with their known position in the grid system. This produces a wrong neatline but the error is unknown to the operator.

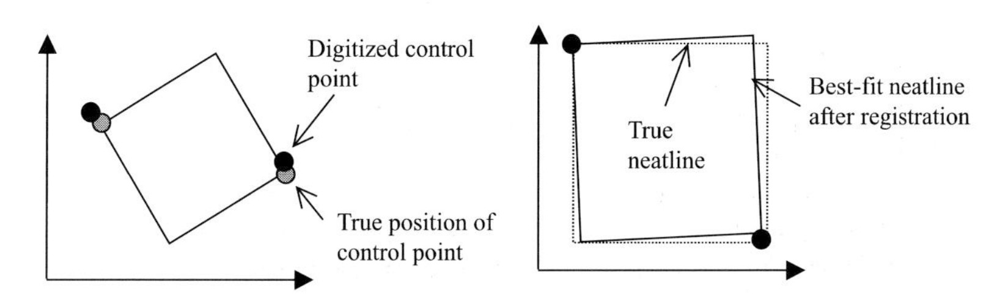

Fig. 5.7 Effect of digitizing errors on registration

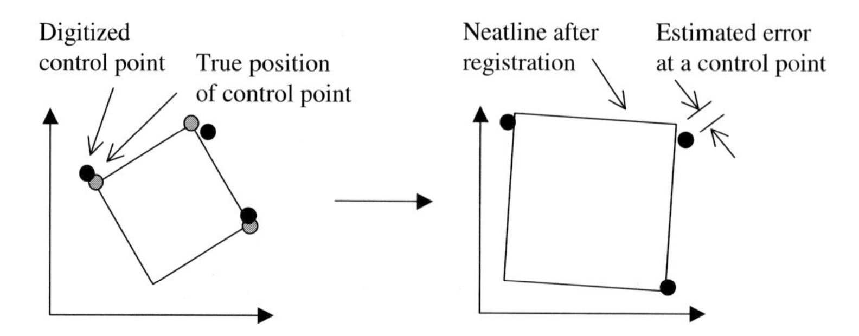

Fig. 5.8 Effect of over-determination on registration

If we digitize more points than are required by a transformation (called *over-determination*), the extra points can help us estimate the digitizing error. As shown in Figure 5.8, we have digitized three points instead of the minimum of two points for a similarity transformation. Since there are errors in the digitized points, it is impossible to find a square that will fit these three points exactly. A usual practice is to find the best-fit solution using the technique of *least-squares adjustment* [Mikhail, 1976]. This will cause the neatline produced to not fall exactly on the control points, allowing us to compute the displacement at each of them. This is an indication of error and is expressed as *residual* by some digitizing software (see Chapter 3). The operator can use the magnitude of these residuals to judge the accuracy of the registration process.

We sometimes register the table coordinate system to fit the geographic coordinate system. That is, we digitize control points on the map and use their geographic coordinates (instead of the usual grid coordinates) in the registration. The similarity, affine, and projective transformations used in registration cannot produce an accurate result. These transformations were designed to transform between two flat surfaces, but in this case we are transforming from a flat table system to an ellipsoidal geographic system. When the area covered is small, however, the error so produced is negligible. Such a transformation is properly handled by map projections, as explained in Chapter 4.

5.4.5.2 Patch-Wise Registration

If the original map has been distorted to the extent that even the projective transformation cannot correct the errors because straight lines have been bent into curved ones (Figure 5.9), then we need to employ *patch-wise registration* methods. These methods divide the entire map into smaller cells, and one of the transformation methods mentioned earlier will be used to register each cell. The assumption is, when the area to be registered is small enough, the effect of distortion will be much reduced. In other words, if the map has been divided into smaller rectangles, a similarity transformation

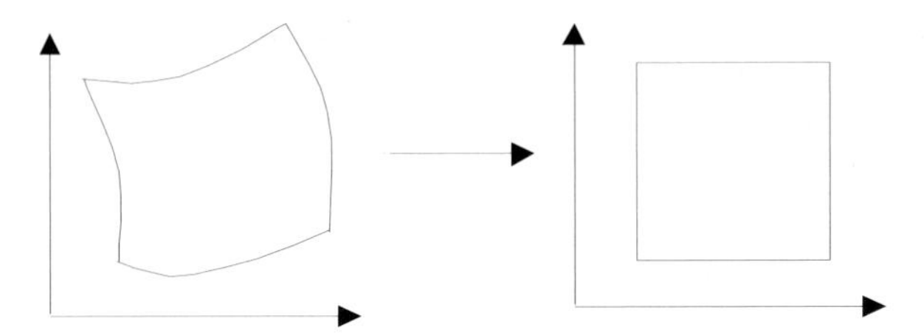

Fig. 5.9 A case for special transformation model

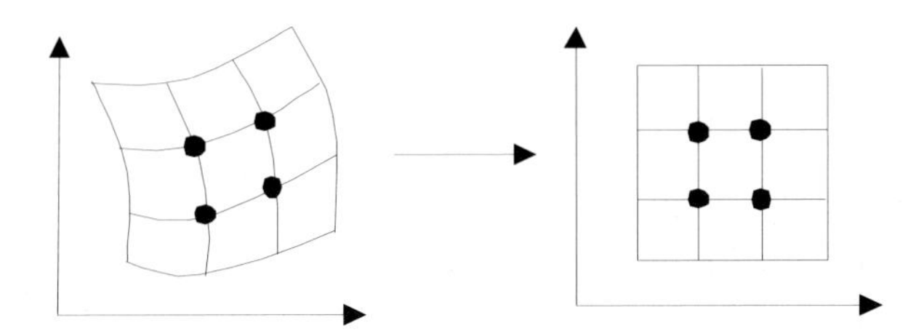

Fig. 5.10 Patch-wise registration on a square grid

could be used to register each rectangular cell and still obtain good results. A common practice is to divide the map according to the rectangular grids on the map because control points within each grid cell are readily available – they are the grid corners (Figure 5.10).

Another method is to divide the map into triangular cells and use an affine transformation for each cell (Figure 5.11). In this case, the three corners of the triangles could be the control points for that cell. This method is not very useful for digitizing a new map because it might not be easy to find so many control points in the right places. When updating a map, however, the problem is to match an existing map on the digitizing table to an existing database. Points that can be identified on the map and also in the database are good control points because their grid coordinates can be retrieved directly from the database.

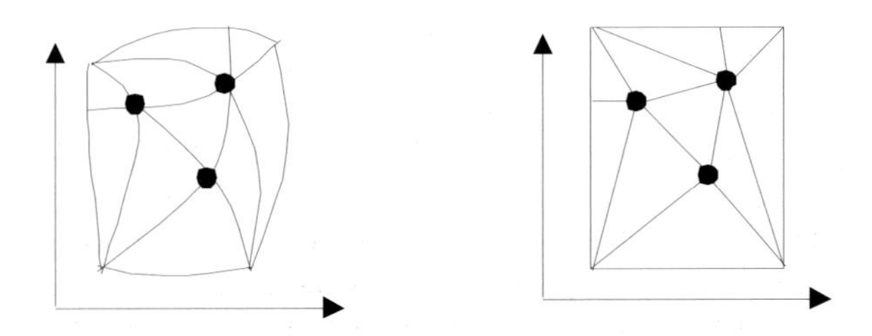

Fig. 5.11 Patch-wise registration on a triangular grid

5.5 Raster Scanning

In a strict sense, raster scanning is also a form of digitization except that the result is a matrix of pixels instead of vectors as explained in section 5.2. For the sake of convenience and to follow the common use of the term, we will use raster scanning (or simply *scanning*) to mean the use of a scanner to convert an analogue map into a matrix of pixels, and we will use line digitizing (or simply *digitizing*) to mean the tracing of lines on a digitizing table to produce a vector map.

5.5.1 Scanner

A scanner converts an analogue document to produce a digital file in raster format. It is an optic-mechanical device that uses a sensor to traverse systematically the surface of the document to be scanned and convert the reflectance (colour or gray level) of its surface to digital signals that will eventually be converted to values of pixels.

Scanners are different in their accuracy, resolution, and cost. They range from the low-cost desktop scanner to the high accuracy drum scanner for commercial applications. There are two aspects of resolution for scanners: spatial and spectral. *Spatial resolution*, similar to resolution for vector, affects the size of the smallest details that can be detected. This affects the precision and fidelity of the product. Naturally, a higher resolution scanner can reproduce finer lines and an output that better resembles the original. The finest resolution may reach 3000 dpi (dots per inch), but in general a resolution of 600 dpi is more than adequate for map digitizing.

Radiometric resolution is the smallest change in reflectance value that can be detected or stored. This is often referred to as the *bit depth* or *colour depth* in scanner terminology, which is the number of bits used to store the reflectance value (gray level or shades of colour) of a pixel. For a bit depth of one, only two values can be recorded to produce a monochromatic image. Eight bits can store 256 shades and is sometimes referred to as indexed colour. Twenty-four bits can store 16.7 million shades and is sometimes referred to as true colour.

Please refer to Chapters 9 and 12 for additional information about raster scanners and the characteristics of scanned images.

5.5.2 Scanning for Cartographic Output

A common use of raster scanning is to convert a line map into a digital format and then display or plot it. Essentially, this makes a "copy" of the original document. In some cases, this copy serves as a backdrop to a vector map display.

To scan a line map for this purpose requires little preparation. Some touchup may be involved either beforehand on the original document or afterwards on the digital version. The only other major considerations are the resolution for scanning and the size of the output.

5.5.3 Raster Registration

Just like points captured from a table digitizer for vector maps, the pixels from a raster map must also be geo-referenced to a grid or geographic coordinate system. The process of registering a raster image is the same as registering a vector map. Control points with known coordinates in the target system (grid or geographic) must be identified from the image. Then an appropriate transformation is applied to the scanned raster image. An additional complication here is a square pixel produced by scanning could transform to an arbitrary quadrilateral after transformation. A re-sampling process would be required to break them up into square pixels again. This will be discussed in more detail in Chapter 10.

5.5.4 Scanning for Data Analysis

If the scanned data is for analyses in raster format, the reflectance value of a black-and-white document is not of much use because it merely indicates the presence or absence of graphics and not the meaning of the underlying geographic features. For a colour document, however, the reflectance value does provide limited attribute information, such as blue lines are most likely hydrologic features. In order to give each pixel a value reflecting a useful characteristic of the underlying feature, an independent *attribute tagging* operation must be carried out. Sometimes it is possible to perform this somewhat automatically if the classification is simple. For example, if we scan a map separately by feature layers, then we can assign a universal attribute (such as "contour") to all features captured from a contour layer. For a topographic map, its content in most cases has already been separately prepared by cartographers, ending up with layers that are combined at printing time. These layers are usually separated by colour and are called *colour separations*. This allows us to scan just the blue layer and assign an attribute of "hydrology" to features from that layer. In other words, if we can separate the content of a document to be scanned beforehand, then we can automatically tag some of the attributes later.

If the map to be scanned cannot be easily separated into layers, it might be possible to use rather sophisticated software to recognize attributes from the symbology of the features. For example, if we know that footpaths are black dashed lines, then we would invoke a programme that knows how to recognize black dashed lines from the raster image and assign attributes accordingly. This is not always a feasible solution, however, because of the complexity of the software involved. If all else fails, attribute tagging would have to be done manually.

5.5.5 Scanning for Vector Output

When a scanned line map has to be used for further spatial analysis, it is more common to convert the raster map to a vector one. This is called raster-to-vector conversion or simply *vectorization*. In this conversion, the pixels obtained from scanning the lines representing various features must be traced or recognised again by software to re-establish the vectors forming the features. If we could perform this conversion automatically, then we would have developed a completely automated method of line extraction.

Much research has been conducted and software developed to handle automated vectorization. The concept is simple but the operation is complex in practice. Different methods have been developed, and a commonly used one is to first vectorize the boundary of a scanned line forming long polygons. For a monochromatic map, this boundary is where a white pixel meets a black one and is relatively easy to detect and vectorize. What remains after this is to find the "skeleton" or centre line of these polygons. The method described here assumes a raster format throughout.

The main problem of vectorization is to automatically recognize a series of pixels forming a continuous line. If the software knows the start pixel of a line (such as the one outlined in white in Figure 5.12), it must decide among its eight neighbouring pixels a continuation. This is not an easy decision. One strategy is for the software to use a skeletonization technique that reduces all lines to a thickness of one pixel (Figure 5.13). If we assume that

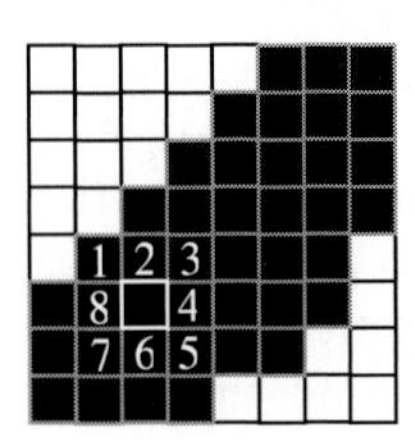

Fig. 5.12 The eight possible connections

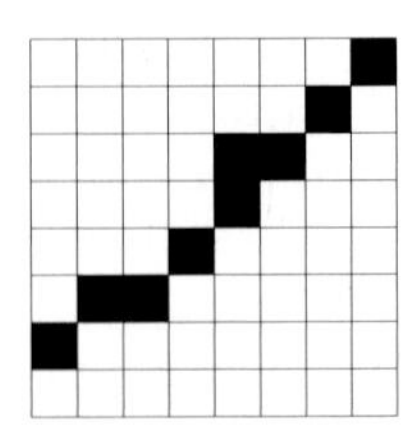

Fig. 5.13 A line of one pixel thin

pixels overlapping graphics are black in colour, this essentially reduces the number of neighbouring black pixels to one (end of a line) or two (middle of a line). This greatly facilitates the tracing operation. For cartographic reasons, the vectorized line could be smoothed to remove the staircase effect.

5.5.5.1 Problems with Automated Vectorization

There are two major problems with the automated conversion of raster to vector maps. One of them is the effect of graphics on line extraction involving both the skeletonization and line tracing operations. The other is attribute tagging.

Line tracing would be easy if all lines were of uniform width, clean, continuous, and non-intersecting. Intersection makes it difficult for the skeletonization to reduce the number of neighbours for each pixel to just one or two. At the intersection, there will be more than two neighbours to a pixel, thus confusing the line tracing operation (Figure 5.14). The same problem happens when lines are not of uniform width. When the document is not clean, the noise created by specks of dirt or ink can bridge lines that are too close to each other and make some lines appear to thicken suddenly. These all cause problems to the skeletonization and line tracing operations.

Lines could become discontinuous when an imperfection on the document breaks the line. Lines could also break due to symbology (Figure 5.15). Although software could be instructed to connect uniform gaps and produce a continuous line from a dashed line, some of the more complex line patterns (such as a cliff) are very difficult to reconcile.

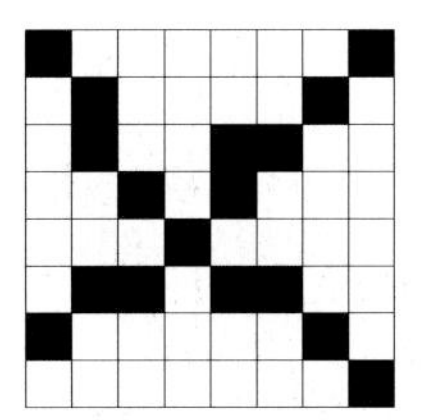

Fig. 5.14 Intersection of two lines

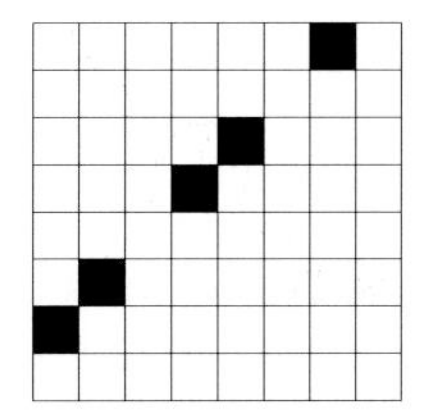

Fig. 5.15 Break of line due to symbolization

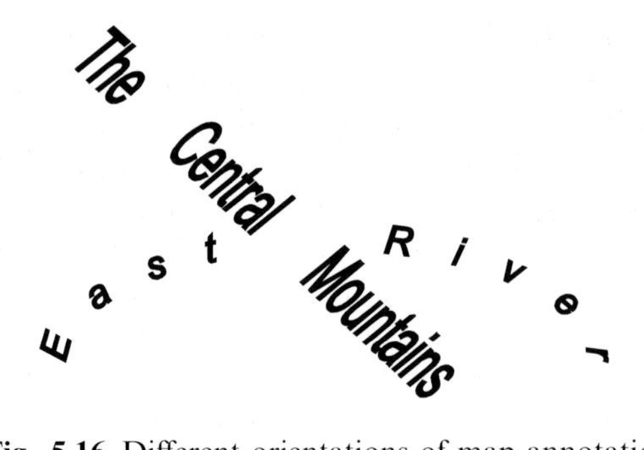

Fig. 5.16 Different orientations of map annotation

A vectorized map would be useless without attributes. As explained in section 5.5.4, it is sometimes possible to recognize and tag attributes automatically. It would be possible to recognize a symbolized point feature and assign the proper attribute to the point. For example, software can be made to recognize a scanned version of a hospital symbol, compute the centre of the symbol, and tag the attribute "Hospital" to the "vectorized" point. This could be done by comparing the scanned symbols with a predefined set of templates of known symbols.

The same technique could be applied to the recognition of text to form names, another important source of attribute information from maps. Automated text recognition is a well-developed technology, but the traditional one used for block text cannot be applied directly to map annotations. For one thing, map annotations do not follow a uniform orientation (Figure 5.16). Rotated characters cannot be matched with the stored template without first restoring them to their normal orientations. Since many names follow the curves of rivers and highways, one must first find the curvilinear feature a name binds to and use it to correct the rotations of the characters.

After recognizing individual characters, we still have to string them together to form a complete name. This can be rather difficult for map annotations because it is not always easy for software to determine accurately the start and end of a name, particularly when annotations can cross each other on a map.

Most of the other attributes, however, cannot be recognized automatically but must be tagged manually. To sum up, completely automated raster-to-vector conversion together with attribute tagging is possible only for simple cases.

5.5.5.2 Semi-Automated Vectorization

We have pointed out in the section 5.5.5.1 that considerable editing must still be carried out after vectorization to make the product useful. Another strategy is to let the operator help the vectorization software interactively to extract lines and to tag attributes.

Earlier scanners for semi-automated vectorization use an optical sensor to detect the presence of lines on hardcopy maps. Line following is done by the sensor performing a local search, usually within a circular neighbourhood, to determine the direction to follow. These kind of line-following scanners operate in a manner very similar to a human operator tracing a line manually. These high-cost devices are gradually being replaced by completely software-driven versions working on a rasterized map.

These soft-copy systems require that the map be scanned and the raster image skeletonized as before. Then, the operator inspects the image and identifies the start and stop (usually at intersections) of a line to be traced automatically by the software. When the tracing software encounters difficulties, the operator can intervene. If the operator thinks the condition of a line would confuse line tracing, the operator can choose to digitize the line manually on the screen. This will at least save one checking step at the end of a completely automated process and could save considerable editing effort afterwards. Annotating and attribute tagging are mostly done manually in this case. A comparison between manual digitizing and semi-automated vectorization is given in Derenyi et al. [1993].

5.6 Conclusion

There have often been long debates about the merits of vector or raster formats for capturing information from maps. In fact, there is no absolute advantage or disadvantage for either model. Any decision has to be based on the application or operational requirements, and the type and amount of resources available. Users often have to make compromises with minimum data storage, ease of data retrieval and analyses, and maximum degree of resolution and accuracy.

As a general rule, a line map in raster format is more useful for cartographic output or backdrop than for analysis. Developments in conversion between the two types have provided the flexibility of use in either format. More and more GIS software accepts both formats but often regards one of them as the prime structure.

This chapter has touched on the fundamentals of converting analogue map information into digital form. It should be noted that whatever the type of conversion, some information loss and distortion of geometry are inevitable. A thorough understanding of these drawbacks will help capture the data in a more useful way.

6 Ground-Based Positioning Techniques

Steve Y. W. Lam and Yong-Qi Chen

6.1 Introduction

The positions of objects, which include topographical features and man-made structures on and below Earth's surface or under water, can be determined by various techniques. These techniques include photogrammetry and remote sensing, satellite positioning, hydrographic surveys, and terrestrial survey techniques (also called ground-based surveys). The first three techniques are discussed separately in later chapters. This chapter presents only ground surveying techniques, i.e., the measurement of angle, azimuth, distance, and height difference to obtain the position of an object using equipment placed at ground level. To inter-relate all the position information collected by different techniques or from different locations, this information must be referred to a common coordinate system. There are several coordinate systems in use, some of which are globally adopted, while others are locally defined. Their definitions and transformations are discussed in Chapters 2, 3, and 4.

The results of ground surveys are usually depicted and drafted on survey plans or maps with coordinates of points and lines being stored in digital format inside a database so that they can be used in engineering design, town planning, and other applications. In the construction of an engineering project, ground surveying plays an important role, including the setting out of the components of the project and quality control during its construction, and monitoring the deformations after its completion. These will not be discussed in this chapter, for the book focuses on data acquisition for geographical information.

This chapter first discusses the measurement of angles and distances and how they are used to determine planimetric information. This is followed by the measurement of height differences and the tying of coordinates measured in the field to a larger geodetic control network.

6.2 Measurement of Distances and Angles

The measurement of angles and distances is central to surveying activities. From these observations it is possible to derive the relative positions and

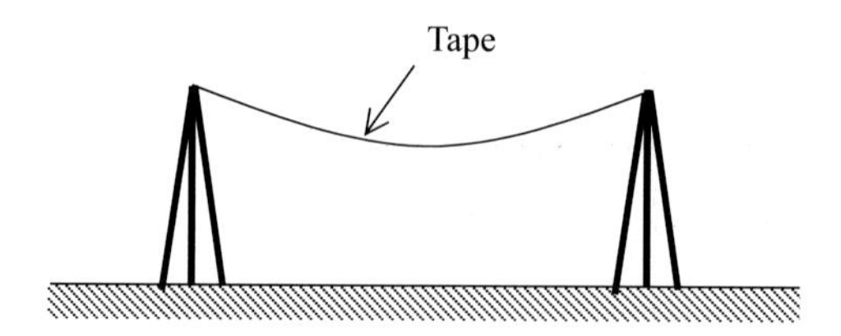

Fig. 6.1 A tape suspended in catenary

elevations of points. Depending on the precision and accuracy required of the survey and the working environment, the distance between two points can be measured using the following methods.

1. Direct measurement with a tape (steel or nylon), which is supported on the ground or suspended in catenary (Figure 6.1). Millimetre accuracy can be achieved by applying proper corrections for standard, catenary, tension, temperature and slope to the measured distance.
2. Electronic Distance Meter (EDM as shown in Figure 6.3a) that utilizes infra-red or near-laser light waves to measure a distance through the reflection of an optical reflector (Figure 6.3b) or directly from the object to be measured. In the reflectorless case, the accuracy of the distance measured depends on the reflectivity of the surface. In both cases, distance is calculated from the modulation frequency and the travel time of the light wave to and from the target. Due to the effect of the atmosphere on the propagation of light waves, atmospheric corrections are required for high accuracy. The achievable accuracy varies from $\pm(10\,\text{mm} + 5\,\text{ppm})$ to $\pm(1\,\text{mm} + 1\,\text{ppm})$ depending on the model of the EDM being used. The first part of the specification is independent of the distance measured. The second part of the specification is a parts-per-million term that is proportional to the distance measured. For example, if the specification of an EDM is $\pm(10\,\text{mm} + 5\,\text{ppm})$, the standard deviation for measuring 2000.000 m by the instrument will be ± 14.1 mm given by this equation:
$$\sqrt{10^2 + \left(5 \times \frac{2000000}{1000000}\right)^2}.$$
3. Optical distance-measurement methods that determine the distance between two points by angular observation on a graduated staff. With the wide use of EDM, such methods have lost their popularity.

An angle is defined as the difference in direction between two straight lines on a horizontal plane (known as a horizontal angle), or on a vertical plane (known as a vertical angle) (Figure 6.2). If a horizontal angle is measured from the grid north, which is along a line parallel to the north-axis of a rectangular grid system (Chapters 2 and 4), it is called a grid azimuth or whole circle bearing (WCB).

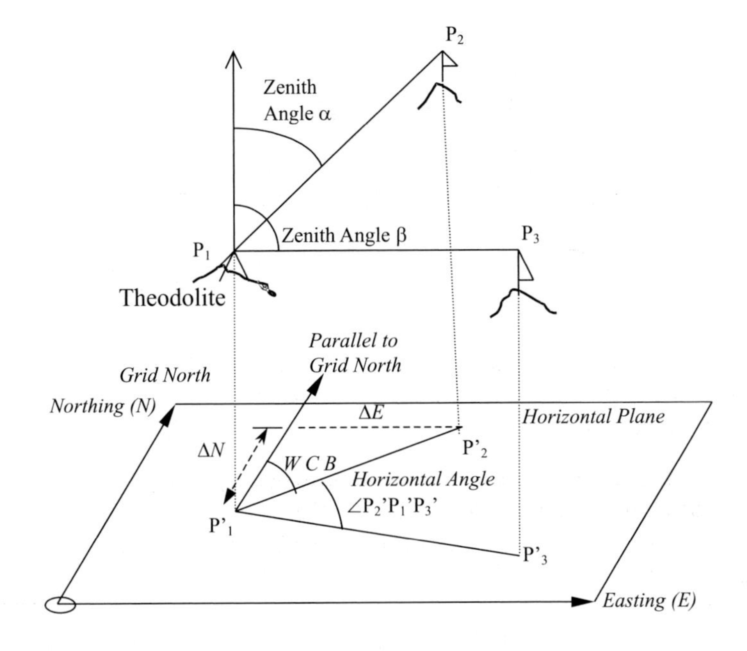

Fig. 6.2 Relationship of angles, distances and coordinates

Horizontal angles can be measured by an optical-mechanical theodolite, an electronic theodolite (Figure 6.3a), or a total station (Figure 6.3c). Theodolites, also called transits, are used in surveying to measure horizontal angles on horizontal planes and vertical angles on vertical planes. They are now gradually being superceded by total stations, which uses a single instrument to measure distance and direction together with data recording.

For a vertical angle, it may be measured to a point above the level line (known as the angle of elevation) or below the level line (known as the angle of depression). For most theodolites and total stations that are used in measuring angles in surveying, a vertical angle is measured from the zenith, which coincides with the vertical axis of the instrument but in the opposite direction above the centre of the instrument, and this is called the zenith angle.

In most countries, the sexagesimal unit is adopted for angular measurements and angle measuring devices. The sexagesimal system uses angular notation in increments of 60 by dividing a circle into 360 degrees with 60 minutes in one degree and 60 seconds in one minute. An azimuth or whole circle bearing is such an angle being reckoned from grid north, around east, south, and west, to 360° with east being 90°, south 180°, west 270° and north 360° or 0°.

In the following example, we will illustrate the basic principle of ground surveying, which is to obtain the coordinates of an unknown point through

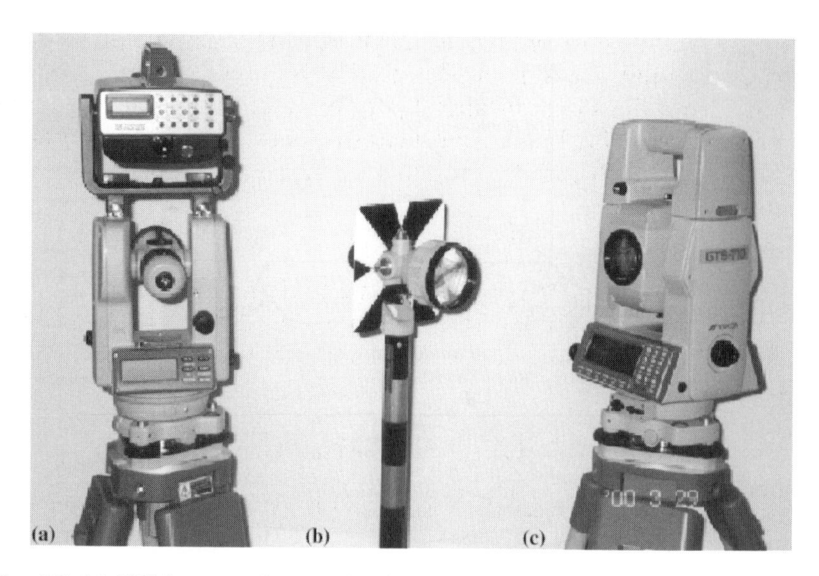

Fig. 6.3 (a) EDM mounted on a Theodolite, (b) Reflector, (c) An Electronic Total Station (Courtesy of Topcon Corporation)

observations on the ground and with the help of other points with known coordinates. On the ground, a series of angular and distance measurements will be made resulting in measurements in the polar coordinate system, as explained in Chapter 2. Note that in ground surveying, angular measurements are mostly from the grid north and calculations are based on the WCB, as explained earlier in this section. Grid north, as explained in Chapter 4, is the north direction implied by the rectangular grids on a certain map projection. After all, the main objective of ground surveying is to determine the positions of ground features on a horizontal plane defined by a particular map projection.

Suppose we know the two-dimensional coordinates of P_1 and P_2 and would like to determine the coordinates of P_3. This can be achieved by making observations to P_2 and P_3 from P_1. The reason for observing P_2 is to compute WCB_{1-2}, the horizontal bearing from P_1 to P_2. Another observation to P_3 will yield the horizontal distance to P_3 and the horizontal angle $\angle P'_2P'_1P'_3$, allowing us to finally obtain WCB_{1-3}, which is the sum of WCB_{1-2} and $\angle P'_2P'_1P'_3$. After this, the horizontal coordinates of P_3 can be calculated by converting from a polar coordinate system to a Cartesian coordinate system, as explained in Chapter 2.

We will now show the detailed calculations involved converting polar coordinates to Cartesian coordinates. Following the surveying and mapping convention, the horizontal coordinates of a point i are expressed in terms of northing (N_{Pi}) and easting (E_{Pi}). Northing is a coordinate value in the north (Y) direction, whereas easting is a value in the east (X) direction. These northing and easting coordinates are dependent on the map projection used to represent Earth's ellipsoidal surface on a plane (Chapter 4).

In Figure 6.2, the horizontal angle $\angle P'_2P'_1P'_3$ can be measured directly by a theodolite or total station in the field. Horizontal distance $P'_1P'_3$ is determined from the slope distance P_1P_3 and the zenith angle α measured by an instrument at P_1. If the coordinates of points P_1 and P_2 are known, WCB_{1-2} from P_1 to P_2 can be computed by the following formula:

$$WCB_{1-2} = \tan^{-1}[(E_{P2} - E_{P1})/(N_{P2} - N_{P1})]$$

where N_i and E_i are the northing and easting coordinates of point i. It should be noted that WCB is always a clockwise angle measured from the grid north with a magnitude between 0 degree and 360 degrees. Thus, WCB_{1-3} from P_1 to P_3 is given by:

$$WCB_{1-3} = WCB_{1-2} + \text{horizontal angle } \angle P'_2P'_1P'_3$$

and the coordinates of P_3 can be computed from coordinates of P_1 by the following formulae:

$$\begin{aligned} N_{P3} &= N_{P1} + \Delta N \\ &= N_{P1} + ((P'_1P'_3)\cos(WCB_{1-3})) \\ &= N_{P1} + ((P_1P_3)\sin(\beta)\cos(WCB_{1-3})) \\ E_{P3} &= E_{P1} + \Delta E \\ &= E_{P1} + ((P'_1P'_3)\sin(WCB_{1-3})) \\ &= E_{P1} + ((P_1P_3)\sin(\beta)\sin(WCB_{1-3})). \end{aligned}$$

To make observations in the field using a surveying instrument such as a total station, we need to position it directly on a point and level it so that measurements are truly referenced to a horizontal plane. Even for measuring vertical angles, we need to reference them to the zenith, which is vertical to a horizontal plane. To place the instrument directly over a point, usually a control point with known coordinates that is physically marked on the ground, we seek the help of a plumb bob. A simple plumb bob is just a heavy weight with a downward point attached to a string. An optical version of it does the same job in a more elegant way. Gravity will pull the plumb bob along a line vertical to the geoid surface (Chapter 2) along the line towards the zenith. Surveying instruments are so constructed that when the sharp point of a plumb bob is placed over a point on the ground, all measurements will be relative to that point. To level an instrument, we use a bubble level that senses the equipotential surface of the geoid. After centring and levelling an instrument, its rotations will be on the true vertical and horizontal planes at a specific point.

Sometimes P_2 in the above case is not available to obtain WCB_{1-2}, the reference bearing needed to calculate WCB_{1-3}. A gyrotheodolite, which is a theodolite integrated with a north-seeking gyroscope (Figure 6.4), can be used to determine the azimuth directly. It should be noted, however, that the gyroscope yields the gyroazimuth A_G measured from the true north and not the required azimuth from the grid north (Figure 6.4). As explained in

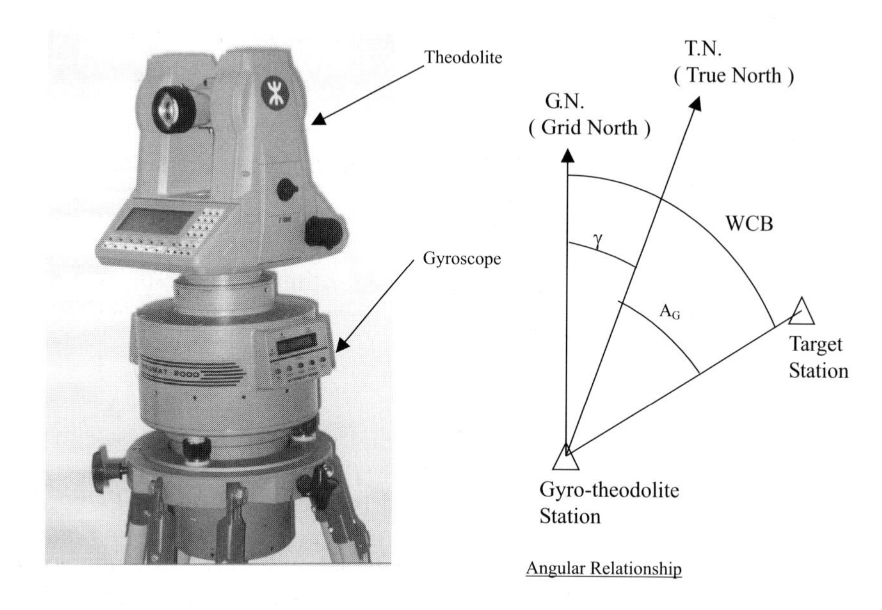

Fig. 6.4 DMT Gyromat 2000 Gyro-theodolite (Courtesy of Leica Geosystems Limited)

Chapter 4, the relationship between the true north and grid north, called the grid convergence γ, is a characteristic of the map projection used and changes from place to place in a systematic way. However, the grid convergence on a map projection can be determined precisely for any point given its longitude and latitude using the following formula:

$$\gamma'' = (\lambda - \lambda_0)'' \sin(\phi)$$

where $(\lambda - \lambda_0)''$ is the difference (in seconds) in longitude from the central meridian λ_0 of the origin of the grid rectangular system, and ϕ and λ are the latitude and longitude respectively of the gyrostation.

After knowing the grid convergence at the station, we can then convert the gyroazimuth to WCB by:

$$\mathrm{WCB} = \mathrm{A_G} + \gamma.$$

6.3 Measurement of Height

The methods described above do not measure the differences in elevation between points. To find the difference in elevation, or height, we need an instrument that will provide us with a horizontal line of sight. Such an instrument is called a level.

If we know the height difference between two points and also know the absolute elevation of one of these points, the absolute elevation of the other

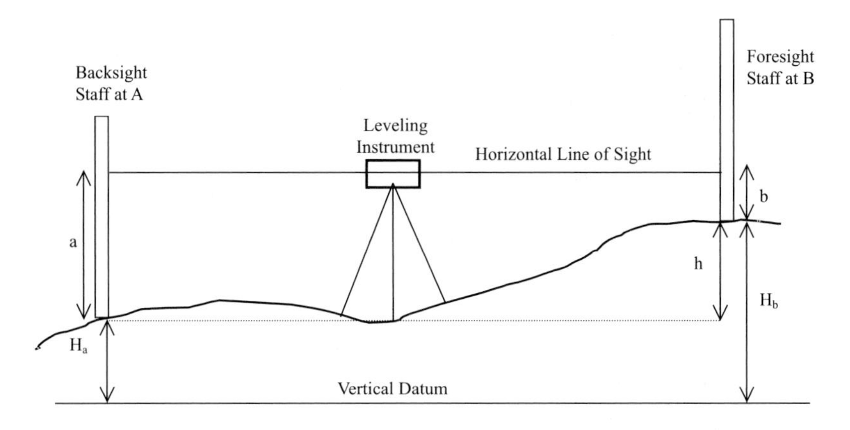

Fig. 6.5 Differential levelling between two points at a level setup

point can then be obtained. As explained in Chapter 2, the elevation of a point is measured from a vertical datum such as the mean sea level (Figure 6.5).

To find the difference in elevation between points A and B, we place a level mid-way between them (Figure 6.5). By levelling the instrument, we can ensure that it rotates on a horizontal plane. Now place a graduated staff, basically a ruler held vertically, at point A and look at it through the level. The line of sight will intersect a marking on the graduated staff. Let this reading be a. Place the staff on B and obtain a reading b on it through the level. Because both lines of sight are horizontal, the vertical distance between A and B is the difference between a and b. This process of surveying is known as differential levelling or simply levelling. As shown in the Figure 6.5, the height difference between A and B is given by:

$$h = a - b$$

where a = backsight staff reading, and b = foresight staff reading.

If two points A and B are quite far apart or not intervisible, several setups are required to determine their height difference. Let a_i be the backsight reading at setup i and b_i the foresight reading, see Figure 6.6. The height difference between A and B is given by:

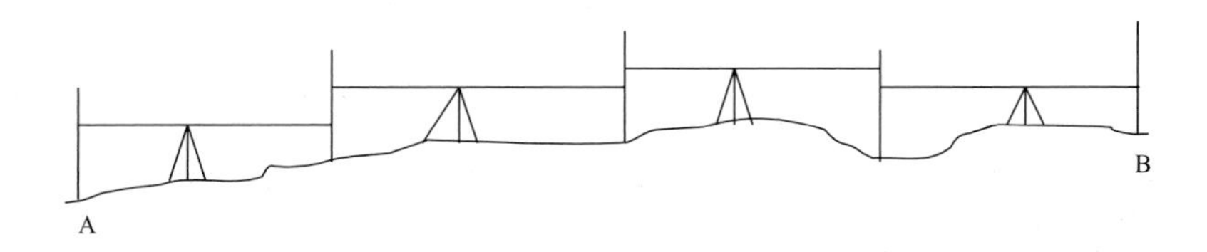

Fig. 6.6 Differential levelling between two points with successive setups

$$h = (a_1 - b_1) + (a_2 - b_2) + \cdots + (a_n - b_n)$$
$$= \sum_{i=1 \text{ to } n} (a_i - b_i)$$
$$= \sum_{i=1 \text{ to } n} (a_i) - \sum_{i=1 \text{ to } n} (b_i)$$
$$= \text{sum of backsight readings} - \text{sum of foresight readings.}$$

An example of differential levelling is shown in Figure 6.7 in which

$$h = (4.380 - 2.175) + (3.565 - 4.230) + (2.210 - 5.580) = -1.830\,\text{m}.$$

Given the height of benchmark AP5 (also called reduced level), the reduced level of Bench Mark "AP6" is 50.290 – 1.830 or 48.460 m.

The elevation obtained by differential levelling is the height about the geoid, which is called the orthometric height as explained in Chapter 2. In theory, the horizontal planes at the levelling stations are not parallel to each other because of the undulations of the geoid surface. When the area covered by the survey is small, the error caused by this is negligible.

The auto-optical level and the digital level are two different types of level (Figure 6.8) used in levelling. A traditional auto-optical level consists of a telescope mounted on a levelling head. When the instrument is properly levelled, the optical compensator of the instrument will give a horizontal line of sight defined by the intersection of the cross hairs inside the telescope. A digital level, which has a built-in camera and a self-levelling device, has to work with a level rod having a bar-coded graduation. During observations, the camera will record the digital image of the rod together with the three cross haris superimposed over the bar-coded graduations. This image is then compared with the master image of the entire rod in the central processing unit to provide the correct rod reading. In short, a digital level uses pattern recognition techniques to automatically read the graduations on a rod. For this reason, they do not work very well in a dark environment, such as inside tunnels, where auto-optical levels are still the instruments of choice. However, digital levels are becoming popular for their cost-effectiveness, ease of operation, high accuracy, and automatic recording, together with their data reduction and downloading functions.

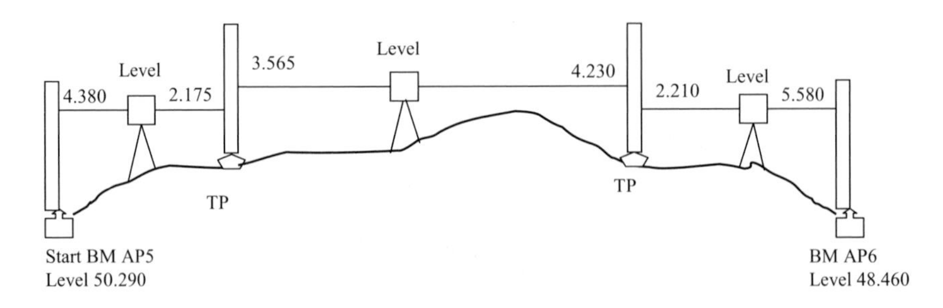

Fig. 6.7 Example of ordinary levelling

Fig. 6.8 Auto-optical level and digital level (Courtesy of Topcon Corporation)

Levelling is a simple but labour intensive and time-consuming operation. If the accuracy requirement is not high, an alternative technique, called trigonometric levelling, can be employed. The technique is also good for the determination of the height of a point that is difficult to access by differential levelling, such as points on rooftops. Trigonometric levelling is the determination of the height difference between two stations by measuring the vertical angles and distance from one point to another (Figure 6.9). In

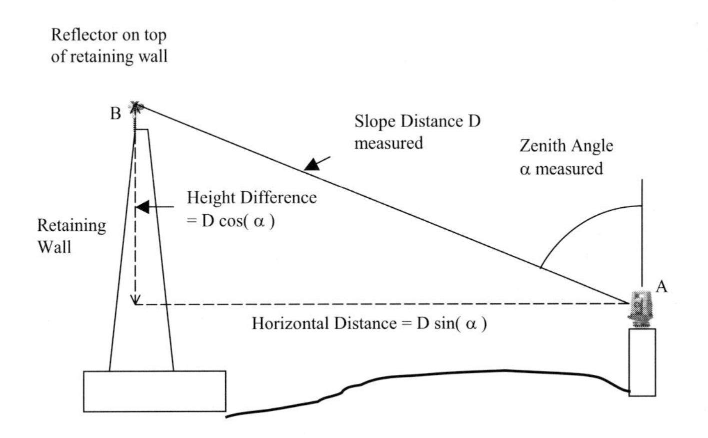

Fig. 6.9 Determination of height difference by slope distance and vertical angle

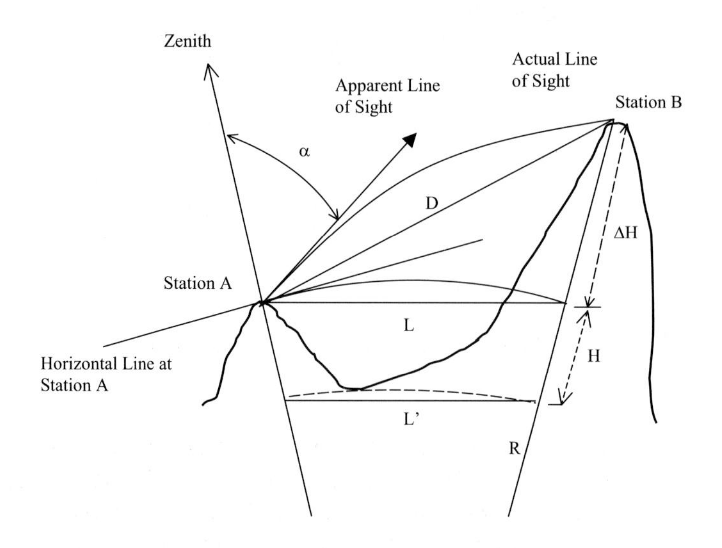

Fig. 6.10 Trigonometric levelling

Figure 6.9, the reduced level of the target centre at B = the reduced level of the instrument at A + D(cos(α)).

Over long distances, the effects of Earth's curvature and atmospheric refraction must be considered in computing the height differences (Figure 6.10). Three corrections should be applied to the observed distance, namely, the curvature correction, the refraction correcton, and the height above mean sea level correction. The height difference ΔH, the horizontal distance L, and the distance L′ on the map projection plane are expressed by the following formula, which assumes a spherical Earth to simplify computations:

$$\text{The height difference, } \Delta H = D\cos(\alpha) + [(1-k)/(2R)][D\sin(\alpha)]^2$$

where D = slope distance corrected for atmospheric conditions
α = zenith angle recorded at the instrument to the target
R = mean radius of Earth (approximately 6370 km)
k = the coefficient of refraction (approximately 0.15).

6.4 Reduction of Coordinates to Datum

Ground surveying using total stations performs measurements and calculations on a level and flat surface, very much the same way that we perform measurements on a plane coordinate system such as a map. There is a major difference, however, between the plane surface used in this case and the flat surface of a map. The flat surface of a map is a plane representation of an ellipsoidal surface, which is an approximation of Earth. The relative position

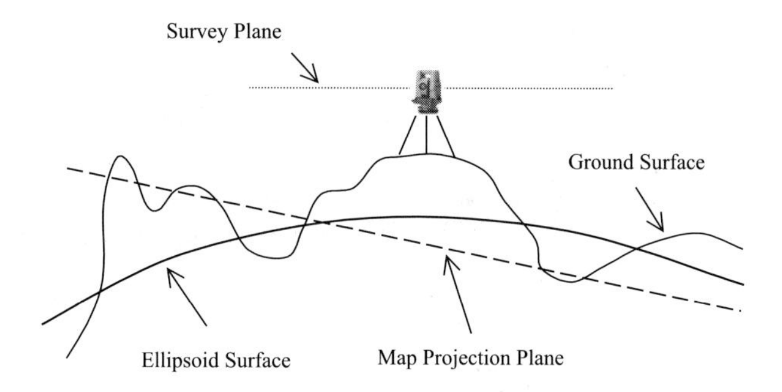

Fig. 6.11 Referenced planes and surfaces in ground surveying

between this surface and the ground is fixed and can be determined if we know the datum (Chapter 2) being used (Figure 6.11). As explained in Chapter 2, all horizontal positions are measured on the ellipsoidal surface. A map projection takes these horizontal positions and transforms them onto a flat surface (Chapter 4). The flat surface of the map could be either tangential to the ellipsoidal earth or cut into it. In general, therefore, the mapping surface is not necessarily tangential to a point on the ground.

The surface used for ground surveying, on the other hand, is normally tangential to the spot on the ground upon which the instrument is located, which can be above or below the ellipsoidal surface (Figure 6.11). After obtaining measurements using a total station, it is necessary to *reduce* the observed coordinates on the ground to those of the datum. Only after such a reduction can these coordinates match those on maps and in geographical databases.

6.5 Detail Surveys

Ground-based surveys are usually either concerned with recording general topographic information, such as the location of roads, rivers, buildings and elevation of the ground, or with thematic information, such as geology, soils, water quality, and vegetation. The latter, more specialized surveys, are based on topographic maps if they exist at an appropriate scale. If they do not, then it may be necessary to carry out some topographic surveying to provide the context for other survey information. Topographic surveys are also referred to as detail surveys.

Detail surveys are conducted based on some control points the positions of which (horizontal coordinates and height) are known. From a control point horizontal angles, distances, and vertical angles to the details close to the control point are measured, and then the coordinates of these details are calculated. Control points are usually well marked or monumented. The heights of control points can be surveyed by differential levelling or

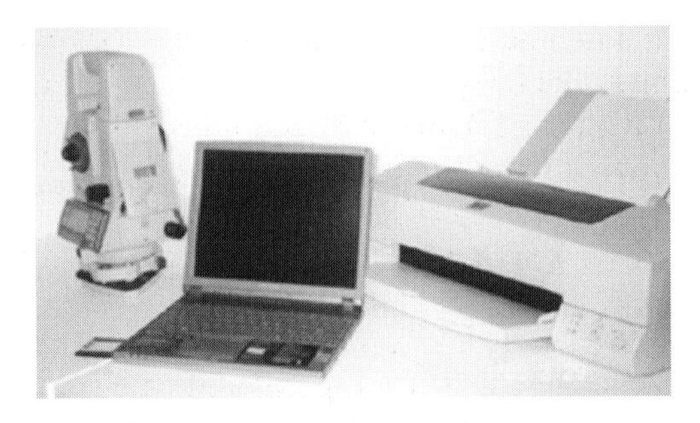

Fig. 6.12 Integration of total station with a computer system

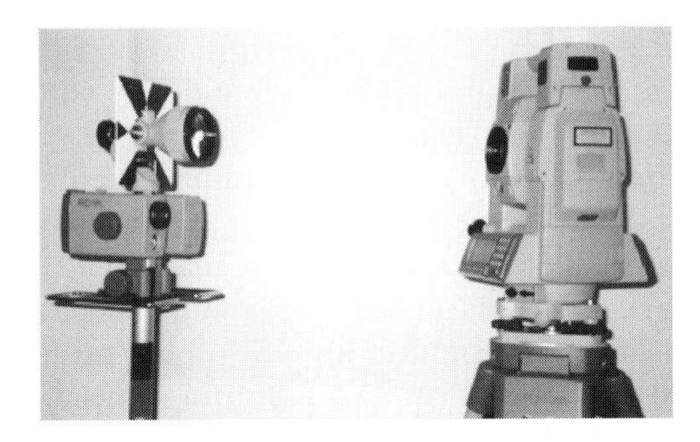

Fig. 6.13 One-person surveying system
(Courtesy of Topcon Corporation)

trigonometric levelling. The horizontal coordinates of control points can be determined with the GPS technique discussed in Chapter 7. Traditionally, the control points are positioned with traversing, triangulation, or trilateration methods. The location of control points depends on the topography or project environment. GPS levelling is now as good as traditional levelling. It can provide centimetre accuracy for relative differential levelling over distances of 50–100 km and decimetre accuracy for distances over 1000 km. GPS levelling is thus applicable in small-scale GIS mapping using handheld GPS devices.

The selection of surveyed points, called sampling, is usually on the basis of critical points relating to the corners of buildings or to changes of slope and ridge lines. This type of sampling is appropriate where the features to be surveyed have an obvious structure that can be characterized by the critical points and lines. Where the distribution is less obviously characterized, sampling may be done on a regular-grid basis.

Total stations become common in detail surveys because both distance and angles can be measured with the coordinates of points computed and displayed automatically by pressing a button on the control panel of the instrument. Each total station can store up to 5000 or more coordinated points inside its internal memory, and has a built-in COGO (Coordinate

Geometry) program to facilitate computations, setting-out, and other surveying operations in the field. Field observation data can be downloaded into personal computers for further spatial analysis and engineering design by a Computer-Aided-Design (CAD) based GIS (Figure 6.12).

One-person total stations are becoming popular for field operations. These one-person robotic systems are made possible by the auto tracking function of a motorized total station and radio communications as shown in Figure 6.13. The system enables one surveyor to carry out the survey, and collect and check data of the survey points by means of the telemetric handheld terminal and palmtop computers. The system also eliminates the need for manual aiming at the prism, and results in fast positioning of points, fewer mistakes, and more efficient surveys in the field.

7 Satellite-Based Positioning

Esmond Mok and Günther Retscher

7.1 Concept of Satellite Positioning

7.1.1 Operation of Satellite Positioning Systems

When satellite positioning is mentioned the assumption is that reference is being made to the American **Na**vigation **S**ystem with **T**iming **a**nd **R**anging, **G**lobal **P**ositioning **S**ystem (NAVSTAR GPS). In fact, there are other similar satellite positioning systems that are under development, for example, the Russian **Glo**bal **Na**vigation **S**atellite **S**ystem (GLONASS). The GLONASS satellites, however, are already available for positioning during the development process. The operation of these satellite positioning systems is very similar, therefore, knowing the concept of one system would lead to an understanding of the others.

The operation of a satellite positioning system is the process of continuous coordination among the ground/control segment, the space segment and the user segment. Each segment has its important role, and the coordination of all these segments makes highly accurate position, velocity, and time determination achievable (Figure 7.1).

Ground-Control Segment

This segment comprises a master control station, transmission stations, and monitor stations. The positions of the monitor stations are accurately known and each has a receiver installed to continuously monitor and receive signals from all satellites in view. The tracking data and meteorological data collected at the monitor stations are then transmitted to the master control station, where the predicted satellite orbit, the health status of the satellites, and their correction parameters are determined, based on the known positions of the monitor stations, to form an important part of the navigation message. The navigation message, formulated at the master control station, is then sent to the transmission stations to be uploaded to all satellites, and broadcast from the satellites to the users.

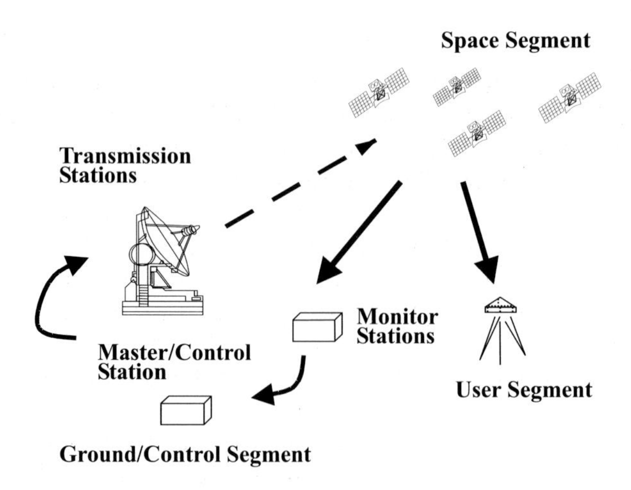

Fig. 7.1 The three segments of GPS

Space Segment

The space segment consists of satellites that continuously broadcast measurement signals and navigation messages to satellite positioning users while orbiting the earth. The satellite configuration varies from one satellite positioning system to another, with the satellites placed approximately 20 000 kilometres above the earth's surface. For the GPS and GLONASS, each satellite continuously transmits two L-band signals propagated in sinusoidal manner, referred to as the L1 and L2 carrier phase signals. The L1 and L2 carrier signals also act as mediums to transmit the series of "0" and "1" code sequences (i.e., PRN pseudo-random noise codes) to users by signal modulation techniques, which can be simply described as the superimposition of measurement information, satellite orbit and messages in terms of "0" and "1" code signals on the carrier signals (Figure 7.2).

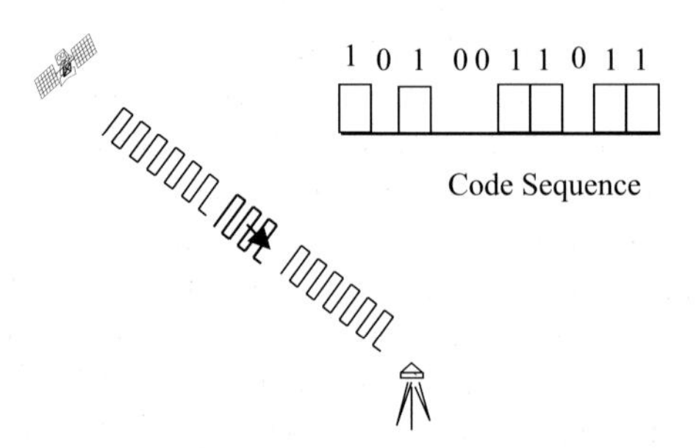

Fig. 7.2 GPS data and navigation messages transmitted to GPS receivers in the form of code sequences "attached" onto L1 and L2 carrier signals

User Segment

This segment can be broadly classified into users for military and civilian applications. There are various types of GPS receivers available on the market to suit different applications, such as surveying and mapping, vehicle navigation, and recreation. Their prices vary from as low as US$100 to as high as US$50 000, depending on whether they have the capability of receiving both L1 and L2 signals, whether they are able to save phase data for real-time or off-site data processing, accuracy specification, and software support. The single-frequency (L1) receiver is normally less expensive than the dual-frequency (L1 and L2) types. The hybrid GPS and GLONASS receivers are also available on the market now but they are more expensive.

A satellite positioning receiver generally contains electronic hardware and software for receiving, decoding, storing, and processing collected data to determine the receiver's position. These data include the navigation message and distance measurements. The orbital parameters of the satellite contained in the navigation message are used to calculate the satellite's position. Distance measurement data derived from codes are called code measurement data, whereas data derived from the sinusoidal phase values of L1 and L2 carrier signals are called carrier phase measurement data. Because the wavelength of the carrier phase data is much shorter than the chip length (i.e. the length between two successive 0 or 1 code signals) of code data, high accuracy position determination from a few millimetres to a few centimetres only can be achieved using the carrier phase data. The use of code data can give positional accuracy of 2 to 5 metres at best.

7.1.2 Satellite Positioning by Distance Measurements

The concept of satellite positioning is not difficult to understand if we first consider a horizontal position located by two horizontal distance measurements. Suppose we want to plot the position of a lamppost on a survey plan. Near the lamppost two building corners A and B as shown in Figure 7.3 can be clearly identified. If the horizontal distance from A to the lamppost is measured and is known to be equal to R_1, then a circle with the building corner A as centre and radius R_1 can be drawn on the plan. The lamppost should lie somewhere on the circumference of this circle. Now, another distance from corner B to the lamppost is measured, and another circle can be drawn. The intersection of the two circles reduces the lamppost's possible location to two positions. If we already have some idea about the approximate location of the lamppost, then the exact location can be correctly selected from the two possible positions. This concept is also known as the concept of the Line of Position (LOP). A single positioning observation tells us something about our position, but not enough to determine a complete position fix. In the two-dimensional case at least a second LOP is required to obtain the position fix. Observations can be

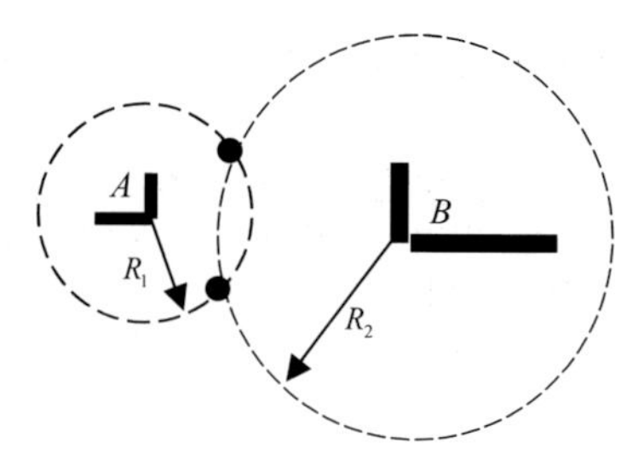

Fig. 7.3 Horizontal position of the lamppost located by two distances

distance measurements, azimuths, distance differences, or angles (or azimuth differences). If we have more than the minimum number of LOPs required, there will be inconsistencies among the intersecting pairs of LOPs. Position determination has to be done then with a least-squares adjustment to account for the redundant observations. The three-dimensional equivalent is the concept of the Surface of Position (SOP). In this case, at least three SOPs are required for position determination.

The concept of satellite positioning is similar to the SOP concept. Let us consider a point *P* above the earth's surface is to be determined by navigation satellites. If at a particular time, a satellite's three-dimensional position is accurately known, and the distance from the satellite to *P* is also known, then a sphere with the satellite fixed at the centre puts point *P*'s position somewhere on the surface of this sphere. If the distances from *P* to two satellites are known, then the intersection of two spheres is a circle, which implies that point *P* is located somewhere within this circle. Following this concept, the intersection of three spheres from three satellites will reduce the possible location of *P* to two points, where its correct position can be identified based on the approximate position (see Figure 7.4). In the case of satellite positioning it is also useful to know that one position will lie on or near the earth's surface and the second may be in outer space [Wells, 1987].

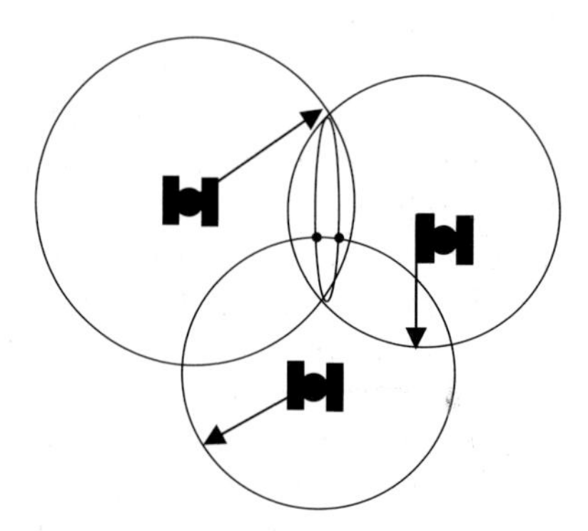

Fig. 7.4 Concept of satellite positioning by Surface of Position (SOP)

7.1.3 How Distance is Determined for Satellite Positioning

There are two ways to define a satellite-receiver range. Measurements using code will generate range information in linear units, namely, range or code range measurement. Another type of measurement, called phase measurement, expresses distance in terms of the number of cycles of carrier wave propagated.

Code Range Measurement

For this type of measurement, distance is determined at the receiver by counting the time delay for a sequence of "0" and "1" codes transmitted from the satellite and the identical sequence of codes arriving at the receiver. The time delay (dt) is multiplied by the velocity of light to obtain the range value. To determine the time delay, a replica of the code sequence is generated within the receiver. This generated sequence is compared with the identical incoming code sequence. The two identical sequences will not align with each other because of the delay caused by the variation of signal velocity when propagating through the atmosphere and other sources of errors. The time shift required to align the two code sequences is the time delay.

The velocity of light is a very large number (299 792 458 metres per second), so even a small error in time delay will cause a significant error in the measured range. For example, a clock error of 1×10^{-6} seconds will result in an error of 300 metres! The main sources of error in the time delay are the satellite and receiver clock errors. Because of the effect of these two errors, the code range measurement data does not give the "true" distance; therefore code range data is also called pseudo-range data.

The measurement of the time delay requires synchronization of the satellite and receiver clocks. Satellites are equipped with extremely accurate atomic clocks. Receivers, however, are supplied with the less accurate quartz crystal clocks because atomic clocks are extremely expensive, inconvenient to use, and impractical for field operation. This creates a problem of how the satellite and receiver clocks relate to each other. The solution is to synchronize both satellite and receiver clocks to a reference time system. In other words, the satellite and receiver clock offsets from a reference time system have to be accurately known. Satellite clock errors can be corrected by applying the correction parameters contained in the navigation message. For the receiver clock error, it is treated as an unknown value and solved together with the three coordinate unknowns during data processing. This is why the use of satellites to determine latitude, longitude, and height (or X, Y, Z coordinates) of a point, requires simultaneous range measurement from at least four satellites. Of course, under some circumstances when the height of the point is already known, three satellites would be sufficient to determine the latitude and longitude of the position.

Carrier Phase Measurement

Let R be the satellite to receiver distance (range) in metres. This distance, if measured by a sinusoidal wave signal with wavelength λ, can be expressed as N cycles of λ plus the fraction of a cycle $\Delta\varphi$, i.e. $R = \lambda \cdot (N + \Delta\varphi)$. It can be seen that phase measurement data contain two components, the integer cycle and the fraction of a cycle, for example (−1 993 673.239). If both components are correct, then the distance R can be correctly determined. In the phase data, however, while the fractional part (0.239 cycle) is accurately measured, the integer cycle part (−1 993 673) is only arbitrarily assigned at the beginning of the satellite tracking process. The exact count of cycles since the beginning is then recorded. Therefore, assuming there are no loss of lock of signals since the beginning of the tracking, the correct distance at time t since the beginning $R(t)$ would be expressed as

$$R(t) = \lambda \cdot (\tilde{N} + \delta N + \Delta N(t) + \Delta\varphi(t)),$$

where

$\tilde{N}$ is the value of the integer cycle assigned at the beginning of the tracking process;
δN is the integer correction to $\tilde{N}$;
$\Delta N(t)$ is the change of cycles at time t since the beginning, this value is accurately recorded;
$\Delta\varphi(t)$ is the fraction of a cycle at time t.

δN is an unknown value in data processing of position. It should be noted that, if there are no loss of lock of signals since the beginning of the tracking process, each satellite would yield one integer correction. If data were collected from four satellites, in addition to the coordinate unknowns, there would be four integer unknowns. The data processing necessary to determine the so-called integer ambiguity is referred to as ambiguity resolution (see section 7.4.2).

7.1.4 Dilution of Precision

Dilution of Precision (DOP) is an indicator to show the strength of satellite geometry in the position and time determination. Good satellite-receiver geometry would have a low DOP, which means a large tetrahedron volume formed from the satellite and receiver positions. Conversely, bad satellite-receiver geometry would cause a high DOP but a small volume. The usually seen DOP indicators are geometrical DOP (GDOP); positional DOP (PDOP); horizontal DOP (HDOP); vertical DOP (VDOP), and time DOP (TDOP). Each type of DOP represents the effect of satellite geometry on the accuracy of horizontal position, height, time, or their combinations (Table 7.1).

One can use the following formula to estimate the effect of DOP on the positioning accuracy.

Table 7.1 Types of DOP representing the effect of satellite geometry on different components

Dilution of Precision	Components
GDOP	Latitude, Longitude, Height, Time
PDOP	Latitude, Longitude, Height
HDOP	Latitude, Longitude
VDOP	Height
TDOP	Time

$$\sigma_P = \mathrm{DOP} \cdot \sigma_R$$

where

σ_P is the positioning accuracy
σ_R is the range measurement error.

If the VDOP and HDOP during observation are 3.0 and 2.0 respectively, and the error in the range measurement is estimated to be ± 5 m, then the positioning accuracy, according to the formula, is estimated to be ± 15 m in height, and ± 10 m in horizontal position.

7.1.5 Effect of Policies on the Use of Satellite Positioning Services

Although satellite positioning has been proven to be very efficient and effective, users should not overlook the policies governing the use of a satellite positioning service. Any changes of the existing policies may seriously affect the availability of data and the existing achievable accuracy. Because GPS was primarily developed by the U.S. Department of Defence for military applications, two different services are provided. They are:

- the Standard Positioning Service SPS for civilian users and
- the Precise Positioning Service PPS for military or authorized users.

In addition, due to military security considerations, the United States Government has imposed two major policies on the use of the civilian GPS positioning service (SPS), namely Anti-spoofing (AS) and Selective Availability (SA). When AS is activated, a code signal known as P-code, which is used to derive higher precision pseudo-range measurement data, is switched to the Y-code. This Y-code can only be accessed by authorized users in the PPS. The civilian users then only have access to the less precise C/A-code signals. SA deliberately degrades the positioning accuracy by reducing the accuracy of the broadcast satellite orbit (ε technique) and putting errors in the satellite clock (δ technique). With these two degradation techniques, it is expected to have the horizontal position fixing using one receiver and C/A code date less than 100 metres, and 156 metres for height, 95% of the time. Experiments have shown that GLONASS positioning using one receiver and code data yields better accuracy, since at present no SA is applied to

GLONASS signals. By decision of the U.S. President, however, SA has been deactivated with effective of May 1, 2000. Therefore the position accuracy for civilian users are significantly improved, i.e., an increase in accuracy by a factor up to 10.

7.1.6 Main Sources of Errors in Satellite Positioning

The main sources of errors arising in satellite positioning can be classified as satellite-related, receiver-related, signal propagation related errors.

7.1.6.1 Satellite-Related Errors

Orbital Errors

It is necessary to have known positions of the available satellites in order to compute the receiver position. Therefore errors in the satellite orbit will propagate to the derived receiver position. Orbital errors mainly stem from the imperfection of the mathematical models describing the satellite dynamics, plus the errors deliberately imposed on the broadcast satellite orbit information when SA is activated. This type of error is in the range of 5 to 25 metres when there is no effect of SA, and can be increased to 100 metres when SA is activated.

Satellite Clock Error

Although on-board satellites are equipped with extremely high precision atomic clocks for time keeping and signal synchronization, satellite clock error is unavoidable. The behavior of satellite clock error (dT) at an instant of time (t) referenced to time (t_0) can be expressed mathematically as,

$$dT = a_0 + a_1 \cdot (t - t_0) + a_2 \cdot (t - t_0)^2$$

where a_0, a_1 and a_2 are coefficients representing, respectively, the clock offset, clock drift, and the rate of clock drift. These coefficients can be extracted from the broadcast satellite message to calculate dT. When SA is activated, the dithering of the satellite clock will result in errors in the a_1 and a_2 terms, causing a possible error of more than 30 metres in the measured distance.

7.1.6.2 Receiver-Related Errors

Receiver Clock Error

Receivers are supplied with inexpensive quartz crystal clocks; therefore, the receiver clock error is much larger than that of the satellite clocks. The clock error is an unknown quantity varying from one receiver to another, and it is time dependent. To determine its value at a specific time requires mathematical modeling of clock behavior to be included in the receiver position computation process.

Cycle Slip

In the ideal situation, once the receiver has started tracking the in-view satellites and if there is no loss of signal lock, the receiver will keep on receiving and recording carrier phase data with the change in cycles referenced to the same integer unknowns as described in section 7.1.3. In real practice, however, some signals may be obstructed by objects such as high-rise buildings and dense vegetation, or interfered with by other signals. Under these situations, loss of lock of satellite signals will occur. On resumption of lock to these satellites, the previously assigned integer value can no longer be used as referenced for the carrier phase data thereafter. In other words, there will be new sets of integer unknowns to be determined after resumption of signal lock to satellites. This jump of integer value before and after loss of lock is called cycle slip. For correct position determination, cycle slips must be detected and repaired.

Antenna Offset

Distance measurement is supposed to be referred to the electrical centre of the satellite transmitter and that of the receiver antenna. Ideally, the electrical centre of the receiver antenna should coincide with its physical centre. The electrical centre's position is changing from time to time according to the direction of the incoming signals and their strengths. Proper field calibration can be carried out to determine the offset correction. For the position to be determined by two receiver systems with similar signal reception behaviour, orientating the two antennas to the same direction will yield similar offset in direction and magnitude at both stations, thus the between-station differencing technique (section 7.2) can greatly reduce the effect.

7.1.6.3 Signal Propagation-Related Errors

Ionospheric and Tropospheric Delay

Measurement signals in the form of electromagnetic waves are transmitted from about 20 000 km above the earth's surface. The velocity of these signals will change when they propagate through the ionosphere and the neutral atmosphere regions, causing errors in the measured distance.

The ionosphere is a band of atmosphere approximately 50 km to 1000 km above the earth's surface. In this region, free electrons are released as the result of gas molecules being excited by solar radiation. The number of free electrons in a 1-square metre column along the path is called Total Electron Content (TEC). TEC varies from time to time depending on the degree of excitation of the gas molecules. The affecting factors include sunspot activities, and seasonal and daily variations that are location dependent. It should be noted that the propagation of the code signal through this medium would slow down, delaying the arrival of the code signal at the receiver, and eventually causing the measured distance to be longer than the true distance. However, the carrier phase signal propagates faster than the velocity of the modulated signal. This phase advance phenomenon causes the measured signal to be shorter than the true distance. Correction formula can be applied to reduce the ionospheric error. Its effectiveness, however, is largely dependent on the reliability of the estimated TEC. Since the ionospheric error is frequency dependent, a more effective method to reduce this error is to compute receiver positions using "ionospheric free" data derived from two frequency measurements. To do this would require the more expensive dual-frequency receivers.

Tropospheric error refers to the delay of signals when propagating through the atmosphere composed of the troposphere (a layer from the earth's surface to about 15 km), the stratosphere (a layer above the troposphere extending to about 50 km), and the upper atmosphere. The term tropospheric error is used because the troposphere has a dominant effect. Electromagnetic waves propagating through the atmosphere will experience a refraction delay subject to the humidity, temperature, and pressure. Since the frequency band used for satellite positioning has no direct relationship with the delay, the combination of two frequencies cannot remove such an error. Many mathematical formulae have been suggested to reduce tropospheric error, the most popular ones being the Hopfield Model, the Black Model, and the Saastamoinen Model. However, only the dry gas component that constitutes about 90% of the error can be determined quite accurately. The remaining 10% on the modeling of water vapour content is much more difficult to be accurately determined.

Multipath Error

Multipath errors can occur when the measurement data is derived from a mix of direct signal and other unwanted signals reflected from nearby surfaces, such as the sea surface and building structures. The deviation of this combined signal from the direct signal will give rise to errors in the measurement data, which can directly affect the accuracy of the position determination. For code range measurement, multipath error is limited to one chip length, i.e., if one chip length corresponds to 293 m, it is possible to have a 293 m error in distance due to multipathing. For carrier phase measurement, the multipath error does not exceed one-quarter of the wavelength, i.e., 5 cm if the wavelength is 20 cm. Effective ways to reduce this type of error include incorporation of narrow correlator spacing techniques and Multipath Elimination Technology (MET) in the receiver hardware and the use of specially designed antennas.

7.2 Absolute and Relative Positioning

Position can be determined by one or more receivers. 'Absolute,' or 'single-point,' positioning is used to describe positioning using one receiver. Relative positioning requires at least two receivers.

Figure 7.5 shows the concept of absolute positioning. One receiver is used to simultaneously observe the available satellites. The receiver position is determined based on the concept of distance resection from the known orbital coordinates of the satellites. Currently, all satellite positions are referred to the Earth-Fixed-Earth-Centred (EFEC), World Geodetic System 84 (WGS84). Therefore, the computed receiver position has its coordinates defined relative to the origin of the WGS84. Single point positioning is economical and efficient, but the accuracy is low. This is because errors exist in satellite positions, essentially due to SA, as well the ionospheric and tropospheric errors in the distance measurements will directly transfer to the computed solution of the receiver position. Before deactivation of SA (see section 7.1.5), Standard Positioning Service (SPS) GPS users could achieve a horizontal position accuracy less than 100 m 95% of the time with a single receiver and C/A code pseudo-range data. The Precise Positioning Service (PPS) GPS users, i.e., U.S. military and other authorized users, can gain access to both C/A and P codes. Absolute positioning can therefore be achieved within 16 m horizontally and 23 m vertically 95% of the time.

The relative positioning concept is shown in Figure 7.6. One receiver is placed at location A whose WGS84 coordinates are accurately known (X_A, Y_A, Z_A), and the coordinates of B (X_B, Y_B, Z_B) are to be determined. Both

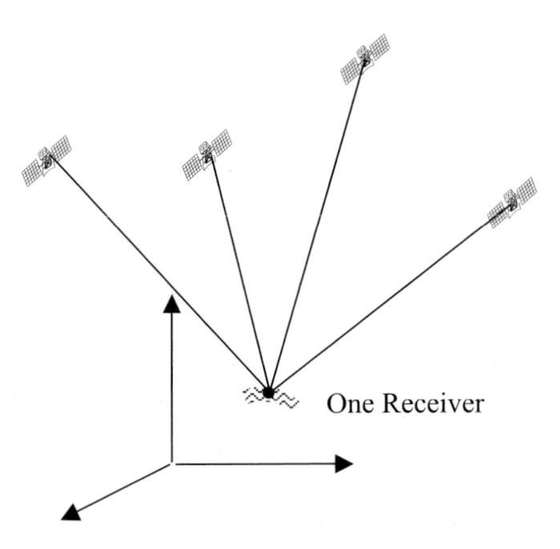

Fig. 7.5 Absolute positioning concept

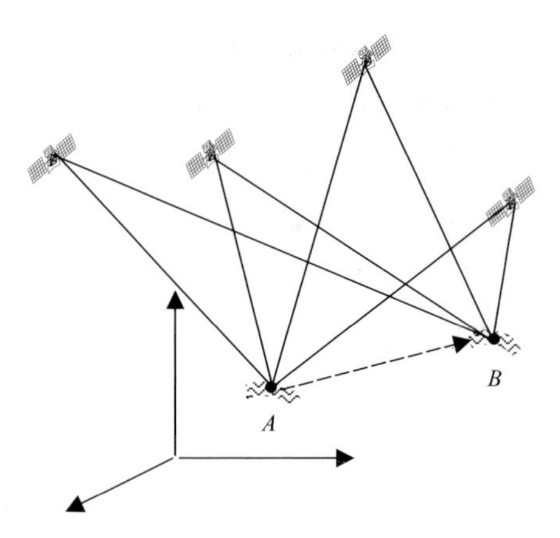

Fig. 7.6 Relative positioning concept

receivers simultaneously observe the common satellites over a period of time, then the change in coordinates from A to B $(\delta X, \delta Y, \delta Z)$ is determined. The absolute coordinates of B are then simply the sum of the absolute coordinates of A and the change in coordinates from A to B, that is

$$X_B = X_A + \delta X$$
$$Y_B = Y_A + \delta Y$$
$$Z_B = Z_A + \delta Z.$$

It will be shown below that errors common to or similar at receiver sites can be eliminated or greatly reduced by 'differencing' techniques. The relative position $(\delta X, \delta Y, \delta Z)$ is then processed by the 'differenced' data where higher degree of accuracy can be obtained.

Let us consider at an instant of time data transmitted from satellites P and Q are collected at receiver stations A and B. In order to make the mathematical expression of the differencing techniques easy to understand, the code range measurement data are assumed. The four measured distances from satellites P and Q to stations A and B can be expressed as their true (correct) distances plus errors as follows:

$$\begin{aligned} R(P,A) &= \rho(P,A) + RC(A) + SC(P) + ORB(P,A) + ION(P,A) \\ &\quad + TROP(P,A) + MP(P,A) + Other\ (P,A) \\ R(Q,A) &= \rho(Q,A) + RC(A) + SC(Q) + ORB(Q,A) + ION(Q,A) \\ &\quad + TROP(Q,A) + MP(Q,A) + Other\ (Q,A) \\ R(P,B) &= \rho(P,B) + RC(B) + SC(P) + ORB(P,B) + ION(P,B) \\ &\quad + TROP(P,B) + MP(P,B) + Other\ (P,B) \\ R(Q,B) &= \rho(P,B) + RC(B) + SC(Q) + ORB(Q,B) + ION(Q,B) \\ &\quad + TROP(Q,B) + MP(Q,B) + Other\ (Q,B) \end{aligned}$$

where

R	is the pseudo-range measurement;
ρ	is the true distance;
RC	is the receiver clock error;
SC	is the satellite clock error;
ORB	is the orbital error;
ION	is the ionospheric error;
$TROP$	is the tropospheric error;
MP	is the multipath error;
$Other$	is other errors that have not been taken into account.

If we subtract the two pseudo-ranges from the two receiver stations to the same satellite, i.e. $R(P,A) - R(P,B)$ and $R(Q,A) - R(Q,B)$, the satellite clock error term $SC(P)$ is cancelled. Let us further consider that the separation between receivers A and B is not too far, say within 10 km, and the height difference between the two stations is not excessive, and that it is likely that the ionospheric and tropospheric effects of the signal transmission path PA and PB are very similar. Hence subtraction of the two ranges will largely reduce these errors. Also, over short distances the magnitude of orbital errors $ORB(P)$ and $ORB(Q)$ will be very similar. Therefore, for a short baseline, the effect of orbital errors can be neglected. The errors arise from multipath and receiver clock errors still exist because these errors are dissimilar at station A and B. Hence, after this 'between-station differencing' technique, the above four 'undifferenced' equations becomes:

$$\begin{aligned} R(P;A,B) &= \rho(P;A,B) + [RC(A) - RC(B)] + [MP(P,A) - MP(P,B)] \\ &\quad + [Other(P,A) - Other\ (P,B)] \\ R(Q;A,B) &= \rho(Q;A,B) + [RC(A) - RC(B)] + [MP(Q,A) - MP(Q,B)] \\ &\quad + [Other\ (Q,A) - Other\ (Q,B)] \end{aligned}$$

Let us rewrite these equations as

$$\begin{aligned} SDR(P;A,B) &= SD\rho(P;A,B) + SDRC(A,B) + SDMP(P;A,B) \\ &\quad + Other\ (P;A,B) \\ SDR(Q;A,B) &= SD\rho(Q;A,B) + SDRC(A,B) + SDMP(Q;A,B) \\ &\quad + Other\ (Q;A,B) \end{aligned}$$

If we further subtract these two single-differenced (SD) pseudo-ranges, it is obvious that the receiver clock errors are eliminated, and the 'double-differenced' (DD) equation is obtained.

$$\begin{aligned} DDR(P,Q;A,B) &= DD\rho(P,Q;A,B) + DDMP(P,Q;A,B) \\ &\quad + Other\ (P,Q;A,B) \end{aligned}$$

It is now clear that, with the differencing techniques, most of the errors can either be eliminated or significantly reduced.

Differencing techniques can be applied to both code range and phase data. However, it is common to use the term 'Differential GPS', or simply DGPS to describe relative positioning using code range data to achieve 3 to 5 metres accuracy. Better accuracy of DGPS can be achieved if the major errors existing in the code range data are filtered with the aid of carrier phase data.

7.3 Differential GPS (DGPS) Technique

The receiver set-up for DGPS is the same as with other relative positioning methods. The reference receiver is set up above station *A*, the position of which is accurately known. The mobile receiver, whether it is stationary or moving, at an instant of time, is situated at *B*. Both receivers collect pseudorange data from at least four common satellites at the same time. In order to achieve higher positioning accuracy, two types of corrections, namely, 'coordinate correction' and 'range correction,' can be applied.

The DGPS technique can be processed in real-time or off-line. For real-time applications, a communication link between the reference receiver and the mobile receiver sites is necessary.

7.3.1 Coordinate Correction

The positions of both the reference and mobile receivers are computed as though single-point positioning was being carried out at the two locations. Since the coordinates of the reference receiver are accurately known, its positioning error at an instant of time can be estimated by comparing the computed coordinates based on range measurements at an instant of time and the known coordinates. If the mobile receiver is not far from the reference receiver, say less than 10 km, the positioning error at the reference receiver side is likely to be similar to that of the mobile receiver. Based on this assumption, the correction estimated at the reference receiver side is applied to the mobile receiver's computed position. This approach, however, has shortcomings. If at a particular epoch of time the two receivers are not observing the same set of satellites, the errors estimated at the reference receiver side cannot correctly reflect the errors at the mobile receiver side. Hence this error correction method, under some circumstances, cannot effectively improve the positioning accuracy.

7.3.2 Range Correction

The range correction can be applied in place of the coordinate correction approach. This is done by subtracting the range measurements collected at the reference receiver from the corresponding true ranges. Questions may arise as to how the true ranges are obtained. Recall that the satellite coordinates at an instant of time can be computed from the broadcast ephemeris, and the reference receiver's position is already accurately known. The true ranges are simply computed from the known satellite and receiver coordinates. The range corrections are then applied to each of the measured ranges observed by the mobile receiver. The corrected ranges are then used to compute the receiver's position by the single-point positioning method. With this approach the problem of inconsistent corrections can be avoided. This method will fail, however, if there are fewer than four common satellites simultaneously observed by both receivers.

7.4 Positioning Techniques for Surveying

7.4.1 Positioning Modes

Positioning techniques for surveying have in common that they are based on differential GPS (DGPS). The types of observable have been discussed in section 7.2. For DGPS either code or carrier phase measurements on at least two stations can be used. The following modes can be distinguished:

- static
- rapid or fast static
- stop-and-go
- reoccupation
- kinematic.

The modes differ mainly in their time span of observation, the time required for ambiguity resolution, and whether the receiver is static, moving, or a combination of the two throughout the observation period.

In the static and rapid static modes two or more GPS receivers receive signals from the same satellites at the same time, whereby one GPS receiver is always placed on a known point. The length of the observation period depends on the number of satellites, the precision requirements, the length of the line, the type of receiver (single or dual frequency), and the atmospheric conditions. Static observations are the most precise and are taken over a time span of from 30 minutes up to several hours. Rapid static observations can range from a few minutes up to 20 minutes. The limiting factor for the rapid static observations is the reliability of resolving the carrier phase ambiguity.

During stop-and-go observations the rover receiver stops on the unknown points for only a few seconds (observation epochs) and collects phase observations. The observation starts on the first rover station that has to be occupied for a few minutes to carry out an initialization process (section 7.4.2). Afterwards the rover is carefully moved to the second station. During the move, the rover receiver has to maintain continuous

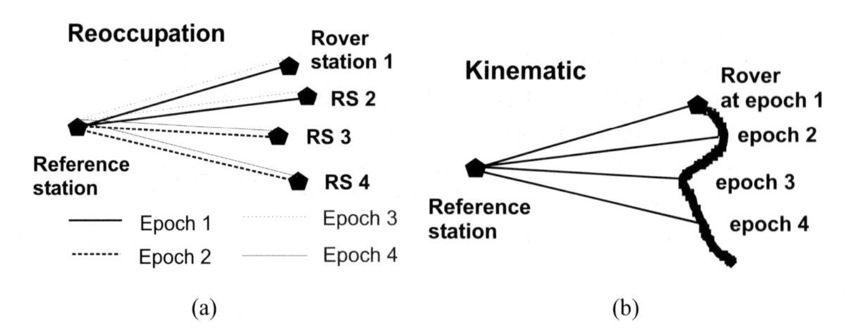

Fig. 7.7 Reoccupation and kinematic positioning modes

observation of at least 4 satellites. Then the observation period at the second station can be reduced to few seconds. If the number of satellites drops below 4, then the rover receiver has to observe again for a longer time period on the next point to resolve the ambiguities or the receiver has to return to a known point (e.g., a previous surveyed point) for re-initialization. The limitations of this method are caused by the requirement of continuous tracking of satellites during receiver motion. Under some circumstances it is not possible to keep tracking the satellites, e.g., the rover receiver has to pass an area with obstruction of the satellite signals or crosses a bridge or goes through a tunnel. In addition, this method is limited to baselines of a few kilometres.

Short observation periods are also used for the reoccupation method. This means that every point is occupied for only a few seconds at two different epochs. During the movement of the rover receiver between the stations, continuous tracking is not required. The time span between two observations on one point should be at least one hour. The processing algorithm is based on the assumption that the satellite-user geometry changes significantly between the two different occupations of a point. Figure 7.7a shows an example of the occupation of four stations. Between epochs 1 and 3 or 2 and 4 a minimum time difference of one hour has to be guaranteed.

All of the above methods require that the rover receiver be stationary for a short period of time. In the kinematic mode (Figure 7.7b) the rover receiver is moving and collects data continuously. Position determination is done separately for each measurement epoch. At the beginning the carrier phase ambiguities have to be determined. This can be done through a static initialization or during the motion (section 7.4.2). After resolution of the ambiguities the receiver has to maintain tracking of at least four satellites for continuous position determination.

7.4.2 Initialization Process

The GPS phase measurement is an "ambiguous" range measurement, and the so-called ambiguity is the initial number of whole carrier-phase cycles at the time epoch of signal lock-on. For precise positioning, this unknown ambiguity must be estimated in the processing step. In the case where a receiver is not stationary for a short period of time, the ambiguity resolution has to be performed using kinematic observations "On-the-Fly" (OTF). The time period from the start of the observation until the ambiguities have been resolved is also referred to as initialization.

Several techniques have been introduced to resolve the carrier phase ambiguities. Initially these techniques involved post processing and required that both the reference and the rover receiver be stationary for a brief period of time during data collection. Ambiguity resolution techniques while the receiver is in motion offer a more or less similar strategy for resolving ambiguities. Basically, the OTF ambiguity resolution is performed by testing many combinations of ambiguity sets inside a certain predetermined search space [Abidin, 1994; De Loach et al., 1995]. Code DGPS solutions can be used to create a search space for the estimation of the correct set of integer ambiguities.

The speed and reliability of the OTF ambiguity resolution are affected by many parameters. To resolve the ambiguities a minimum number of five common observed satellites on both stations under good station-satellite relative geometry (e.g., GDOP < 6) are required. Also, the level of observation errors and biases has to be low, and a high data rate of 1 Hz or greater is necessary. In principle the OTF ambiguity resolution techniques can be applied either for single or dual frequency GPS measurements. In particular, the ionospheric bias degrades the performance of the OTF ambiguity resolution for single frequency data. For real-time applications, nowadays only the more expensive dual frequency GPS systems available on the market provide reliable and fast OTF ambiguity resolution.

Future trends for improving the reliability of ambiguity resolution should make use of combined GLONASS and GPS receivers, which are able to observe more satellites and thus increase the number of observations. A second strategy is based on the recommendation of the U.S. National Academy of Public Administration and the National Research Council to provide an additional frequency, L4, for civilian users in the next generation of GPS satellites [Han and Rizos, 1997].

7.4.3 Comparison of Positioning Accuracy

High accuracies at the centimetre level can be achieved with relative positioning (DGPS). Most commonly static and rapid static observations are performed for precise surveys. Accuracies at the decimetre level can be achieved using the stop-and-go and the kinematic methods (Figure 7.8). Some applications require only accuracy at the decimetre or metre levels and they can be performed using code DGPS. For single-point positioning, accuracies at the metre to 100 m levels are achieved with pseudo-range

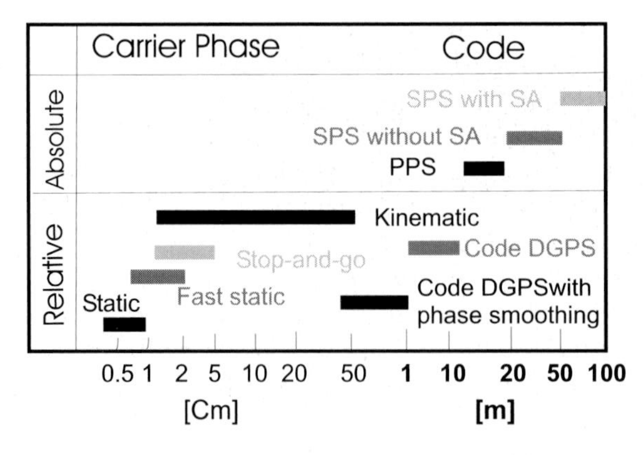

Fig. 7.8 Accuracy spectrum of GPS positioning

observations in the Standard Positioning Service (SPS) and Precise Positioning Service (PPS). These accuracy requirements are typical for any kind of navigation.

7.5 Real-Time Kinematic Positioning

The concept of kinematic positioning was introduced in section 7.4. Conventionally the carrier phase observations are collected during the survey, and the processing is performed afterwards in the office (post processing). Some applications, however, require that positioning determination take place in the field and therefore processing in nearly real-time is necessary. In recent years, GPS receiver systems have been developed that are capable of providing real-time processing. Usually these measurement systems are called real-time kinematic GPS (RTK-GPS). In this chapter the basic principles, requirements, and limitations of these systems are discussed. Figure 7.9 shows examples of RTK-GPS.

Real-time relative positioning with carrier phases is developed from static GPS observations where the receivers are stationary for a certain period of time. The principle of operation is based on relative positioning (section 7.2) where at least two GPS receivers are required. One receiver is positioned on a known point, i.e., the GPS reference station, the others are on the points the coordinates of which have to be determined (Figure 7.6). To distinguish RTK-GPS positioning from code DGPS (section 7.3), the carrier phase observations of the signals L1 or L2 are used for the position determination instead of the code signal. For real-time relative positioning with carrier phases the complete carrier phase data stream has to be transferred from the reference station to the rover station.

Reference station
Geotracer System 3220 RTK

Rover station
(Spectra Precision, Sweden)

Rover station
Trimble Total Station 4800
(Trimble Navigation, U.S.A.)

Fig. 7.9 RTK-GPS systems

7.5.1 Data Link

The phase observations have to be transferred in real-time from the reference station to the rover station. Therefore a telemetric data link is required. In many countries governmental restrictions limit the use of the data link. Special licenses for high transmission powers are difficult to obtain. In Europe and also in Asia, only the frequency bands given in Table 7.2 can be used by private users without requiring a special license. The transmission power of the data link is limited to 0.5 W, which will limit the range of the radio link to a few kilometres.

Three frequency bands can be used which are in the UHF and VHF frequency range (Table 7.2). In general, wider ranges can be obtained using short frequency signals. However, one advantage of the 0.7 m frequency band is that the signal is also transmitted in areas without visual contact to the transmitter due to reflections of the signal, e.g., on a water surface. The range of the data link is affected mainly by the topography of the survey area and therefore no general specifications can be given. Under good conditions it is possible to achieve ranges up to 10 km.

To increase the perceived range of the used HF/MF radio systems either higher transmission powers have to be applied or corrections are broadcasted using other techniques. Alternative techniques are to transmit the

Table 7.2 Specifications of frequencies for radio link

Frequency band	Frequency	Frequency range
4 m	68–87 MHz	VHF
2 m	146–174 MHz	VHF
0.7 m	455–570 MHz	UHF

real-time differential corrections using cellular mobile phone networks or via FM radio signals. The main disadvantage by using mobile phones, however, are the high costs for operation. In the second case, the Radio Data Systems (RDS) standard can be used for the transmission of digital data from existing broadcast FM radio stations. In general, the range of this service is in the order of 50 to 75 km. Another alternative is to broadcast the DGPS corrections using geostationary satellites. Satellite based DGPS systems have been available for a few years and two services now provide worldwide coverage, i.e., Fugro OmniStar and Racal LandStar. Depending on the service, there is a fee or an annual subscription service charge for receiving the corrections [Thomson, 1996].

For RTK-GPS a high data communication rate and a high reliability in the data link is required. Certain format standards have been developed for the data transmission. One is the RTCM (Radio Technical Commission for Maritime Services) format [Kalafus et al., 1986], which has also included OTF differential data message structures in the last version of its recommended format standard. However, for higher data transmission capability and reliability most manufactures of RTK-GPS systems still use their own format standards and therefore the combination of reference and rover stations from different manufactures is yet not possible.

7.5.2 Positioning Accuracy and Applications

The demand for precise horizontal and vertical positioning for survey applications is continuously growing. Starting from code DGPS systems that provide metre to submetre accuracy for position determination, RTK-GPS systems can achieve accuracies at the centimetre level. The systems are either employed in stop-and-go mode, where the rover receiver is stationary for a brief time period (usually a few seconds) and all carrier phase observations are collected, or in kinematic mode, where the position determination has to be performed using on epoch of data (phase observations at a certain time). Higher accuracies can be obtained in the stop-and-go mode depending on the length of the observation period, i.e., ±1 cm up to a few centimetres. In the kinematic mode position accuracies are usually a few centimetres or decimetres depending on the speed of the rover receiver.

Because of the rapid development in instrumentation and techniques a variety of modern equipment is available on the market. Recently new RTK-GPS systems have been introduced by many manufacturers, e.g., Trimble, Leica, Spectra Precision (Geotronics), and Ashtech (Figure 7.9). The systems have in common that similar strategies for the resolution of the integer phase ambiguity on-the-fly (OTF) are applied. Their performance has been analysed for many different applications in surveying. They can be employed for kinematic and static applications where observation periods of a few seconds up to several minutes are required. In addition, using these systems it is now possible to employ GPS for setting out for the first time. Further applications are in the area of precise real-time navigation and position determination.

7.6 Future Developments

The integration of GPS and GLONASS is one step towards future Global Navigation Satellite Systems (GNSS). The GNSS concept is a generalization of the capabilities introduced first by GPS. It combines several components to construct something more comprehensive than GPS and to reduce the reliance on a single system. In its first stage, GPS and GLONASS will be combined with navigation and integrity (i.e., the ability to detect out-of-tolerance conditions) signals from the next generation of communication satellites. Geostationary satellites (e.g., Inmarsat) will provide differential corrections, integrity alerts for "unhealthy" satellite transmissions, and additional ranging signals. In its second stage, the future civil-controlled GNSS will consist of a global network of compatible, regional satellite systems. Strategic goals of the system are to be independent from any military or national system under international control, to guarantee continuity of service over a certain time period, and compared to GPS to offer an improved service which is affordable [Hiller et al., 1997]. At the end of last decade, the European community has also announced to investigate the development of an own global satellite navigation system. This system is called GALILEO and it might be operable by the end of 2008. Due to its system design and similarity to GPS, it can be part of a GNSS together with GPS and other satellite positioning systems.

8 Techniques for Underwater Data Acquisition

Günther Retscher

8.1 Introduction and Overview

Underwater data acquisition is performed to provide information about the sea-bottom. The purpose is to depict the relief of the seabed topography including all natural and man-made features. In general, underwater data acquisition is the major part of a so-called hydrographic survey. The result of this kind of topographic survey is then presented on nautical charts, which are the marine equivalent of topographic maps on land. Their design is fundamentally similar, however, they incorporate additional features that are used for navigation, such as navigation hazards and aids for navigation (leading lights from lighthouses, buoys and beacons, etc.).

Following a definition reached by experts at a United Nations conference in 1979 [United Nations, 1979], hydrography (or hydrographic surveys) can be classified into coastal, offshore, and oceanic. In coastal or inshore waters topographic information of the seabed are required mainly to assure safety of navigation in shallow water. Other purposes of hydrographic surveys in these areas are concerned with the development of ports and harbours, coastal erosion problems, the utilization of harbours, and coastal conservation services. Surveys in these areas are classified as special or first-order surveys (see Table 8.1). Offshore hydrography provides hydrographic data in the region of the continental shelf (water depths up to 200–300 m). Typical applications are the provision of data for laying submarine cables or pipelines, offshore petroleum exploration, and fisheries management. Oceanic hydrography is concerned with the acquisition of hydrographic data in the deep ocean areas for general oceanographic research. Surveys in offshore or deep ocean areas are classified as second or third order.

Nowadays a variety of techniques for underwater data acquisition are available, and a suitable method can be chosen for any type of application. New technologies have dramatically changed hydrographic surveys in the past few years. In general, hydrographic surveys involve two main tasks, i.e., the positioning of a hydrographic survey vessel and the measurement of the water depth below a vertical datum. The maximum positioning and depth errors are specified in accuracy standards established by the International

Table 8.1 Accuracy standards for hydrographic surveys (after IHO, [1997])

ORDER	Special	1st	2nd	3rd
Examples of Typical areas	Harbours, berthing areas, and associated critical channels with minimum underkeel clearances	Harbours, harbour approach channels, recommended tracks and some coastal areas with depths up to 100 m	Areas not described in Special Order and 1st Order, or areas up to 200 m water depth	Offshore areas not described in Special Order, and 1st and 2nd Order
Horizontal Accuracy (95% confidence level)	± >2 m	±5 m+5% of depth	±20 m + 5% of depth	±150 m +5% of depth
Depth Accuracy for reduced depths D_r: $\pm\sqrt{a^2 + (b * D_r)^2}$ (95% confidence level)	$a = 0.25$ m $b = 0.0075$	$a = 0.5$ m $b = 0.013$	$a = 1.0$ m $b = 0.023$	Same as 2nd Order

Hydrographic Organization (IHO, [1997]) and summarized in Table 8.1. The depth accuracy is described with a constant depth error a and depth dependent error b. Depending on the survey area, depth accuracies at the metre to subdecimetre levels have to be achieved at 95% confidence level. Position determination has to be at the metre to 100 m levels.

Figure 8.1 shows the major components of a modern hydrographic surveying system for coastal or inshore waters. Position determination of a survey vessel in the horizontal plane can be carried out with DGPS or electronic positioning systems. For worldwide navigation, radio positioning systems were employed before GPS was available. Nowadays, GPS is used nearly exclusively for positioning at sea. A detailed description of the DGPS positioning technique can be found in section 7.3.

For depth measurements a device called an echo sounder or sonar can be used. The measured depth is referred to as a sounding. Most commonly a survey is carried out in depth profiles, i.e., parallel sounding lines are run. Tidal observations on a wharf or other site are necessary to obtain the current level of the water surface above the vertical datum. Due to tidal variations, time tags are required to synchronize position determination with soundings and tidal observations. The tide record is used to reduce the soundings referring to the datum (reduced depths D_r in Table 8.1). Hydrographic software packages combine the position information with soundings in real-time and store the data on a computer. For on-line navigation on a preplanned sounding line, track guidance features are provided by the software package.

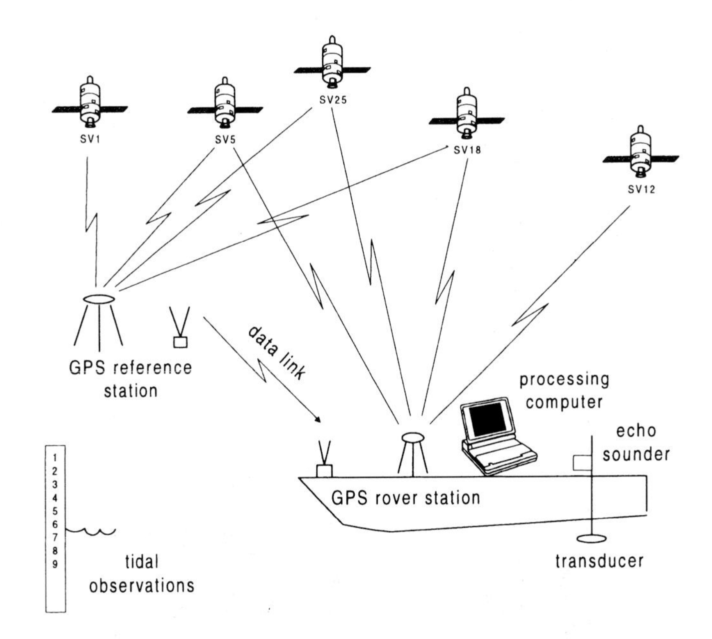

Fig. 8.1 Hydrographic surveying system for coastal or inshore waters

In this chapter, the most commonly used techniques for water depth measurements and tidal observations are described in detail and current developments are explained. It is organized as follows: Water level variations and tidal observations will be described in section 8.2; the techniques for underwater data acquisition will be discussed in section 8.3. Starting with a classification of the types of depth measurements, the operating principles of echo sounders and the physical behaviour of acoustic signals in sea water will be explained. Other techniques will be introduced briefly. Finally, in section 8.4, the steps of operation in a hydrographic survey, the survey preparation, the survey itself, and the data processing, using modern software packages, are discussed.

8.2 Tidal and Other Water Level Changes

The basic theory of tides and the forces producing them is very complex and far beyond the scope of this book. Therefore only some relevant characteristics of tides will be discussed briefly. In order to observe the tide, the height of the sea surface is measured as a function of time with respect to a reference level. Tidal predictions and vertical datum definitions are based on tidal observations. In the following, vertical datums used for surveys on land and at sea are defined, the measurement technologies for tidal observations, and the causes and effects of other water level changes are discussed briefly.

8.2.1 Tidal Characteristics

Tides are horizontal and vertical motions of the sea due to the gravitational attraction of the sun, moon, and other planets. The motions are periodical and are superimposed by random variations. The tidal heights and characteristics at different places vary differently according to their geographical location and meteorological conditions. In general, tides can be classified into four types:

- Semi-diurnal tides with two high waters and low waters every day;
- Semi-diurnal tides with diurnal inequality that have a character intermediate between diurnal and semi-diurnal tides;
- Mixed tides with two high waters and low waters every day when the moon is at the equator, while at other times there is only one high water and low water daily;
- Diurnal tides with one high water and low water every day.

For a certain location on the earth the type of tide is dependent on the geographical latitude and other influences, such as the resonance in a particular ocean basin and the coastline. Theoretically diurnal tides occur in ports at high latitudes and semi-diurnal tides at low latitudes. The tidal ranges achieve maximum values when the gravitational attraction of the two main bodies, the sun and moon, coincide (spring tides) and minimum values when the attractive forces are perpendicular to each other (neap tides). The tidal range varies from over 10 m (e.g. in the Bay of Fundy, Canada) to nothing in a tidal estuary or river. In addition, the tidal range can be increased or decreased by meteorological conditions, such as atmospheric pressure changes and wind (e.g. storm surge) [Wilhelm et al., 1997].

8.2.2 Vertical Datum Definition

A vertical datum is defined based on the analysis of long periods of continuous observations of the sea level. The datum for heights and levels on land is usually referred to as the Principal Datum or Ordnance Datum (see Chapter 2). In many countries geoidal or orthometric heights are used which refer to the geoid as the reference surface. Thereby the geoid is approximated most closely by the global mean sea level. The mean sea level is an average level of the sea surface over an observation period of at least 18.6 years, as this observation period is required to eliminate the main periodic influence of the moon on the tidal variation. In other words, the mean sea level is the average sea level that would exist in the absence of tides. The datum can be defined arbitrarily using the average mean sea level obtained from a certain period of observation as zero mark or in some relation to the average value (see Figure 8.2).

For hydrographic surveys and nautical charts the Chart Datum (CD) is used as the vertical datum. The chart datum is also the datum plane for tidal predictions in Admiralty Tide Tables [Hydrographer of the Navy, 1965–1970]. It should be defined as the zero-depth reference below which the sea level will seldom fall. For most countries, the level of the Lowest Astronomical Tide (L.A.T.) is chosen as the chart datum (Figure 8.2). The L.A.T. is the lowest level that can be predicted to occur under any combination of astronomical conditions and under average meteorological conditions. Depths relating to the L.A.T. datum on nautical charts then show the minimum depth of water available. If the depth value on the chart is, e.g. 5 m, then the water depth will be at least 5 m as the current sea level always will be above the L.A.T. under normal weather conditions. This is very important for the safety of navigation as the actual water depths available for navigation will almost always be greater than that shown on the nautical chart.

As mentioned in the introduction, depths relating to the chart datum are called reduced or charted depths D_r. Then the measured depths D_m refer to the current sea level and have to be reduced to obtain charted depths D_r (see Figure 8.2). A nautical chart shows depths below the chart datum and heights above the Mean Higher High Water (M.H.H.W.). The M.H.H.W. is defined as the mean level of the higher of the two daily high waters over a long period of time.

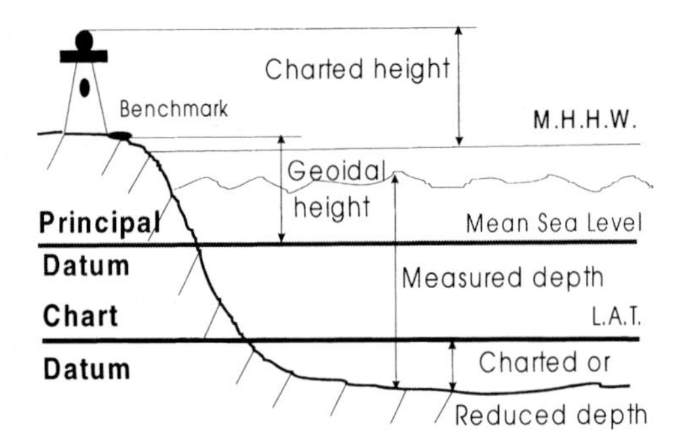

Fig. 8.2 Comparison of vertical datums

If the chart and principal datum of a country are defined as above then the relationship between the two datums is known and very simple. Normally, the chart datum lies a constant factor below the principal datum. The chart datum on a site for tidal observations can be established with levelling from benchmarks referring to the principal datum.

8.2.3 Tidal Observations

Tides can be observed and measured by means of tide poles and automatic recording tide gauges. The simplest method of tidal observation is to use a graduated vertical scale (similar to a levelling staff) on which an observer reads directly the level of water. The pole can easily be erected on a wharf and should cover the total tidal range with its zero at the datum. The main disadvantage of this method is that it requires manual observation and therefore it is very labour intensive. Nowadays, it will be applied only for temporary tidal observations on a site during a small hydrographic survey.

Modern automatic recording tide gauges are used to avoid the disadvantages of manual observation. Depending on the construction, the following types can be distinguished [Hydrographer of the Navy, 1965–1970]:

- Mechanical tide gauges containing a float mechanism where the height of the water surface is measured directly in a stilling well (see Figure 8.3a). A float operates the recording mechanism via a float pulley.
- Acoustic gauges use an acoustic sensor to measure the distance vertically upwards through the water or down through a tube to the water surface (Figure 8.3b). The principle of the range measurement with acoustic signals will be discussed in section 8.3.2.1.
- Pressure operated tide gauges use compressed air to measure the back-pressure of the rising and falling tide either directly or indirectly at a point under the water (Figure 8.3c). The changes in pressure will then be converted to changes in tidal height.

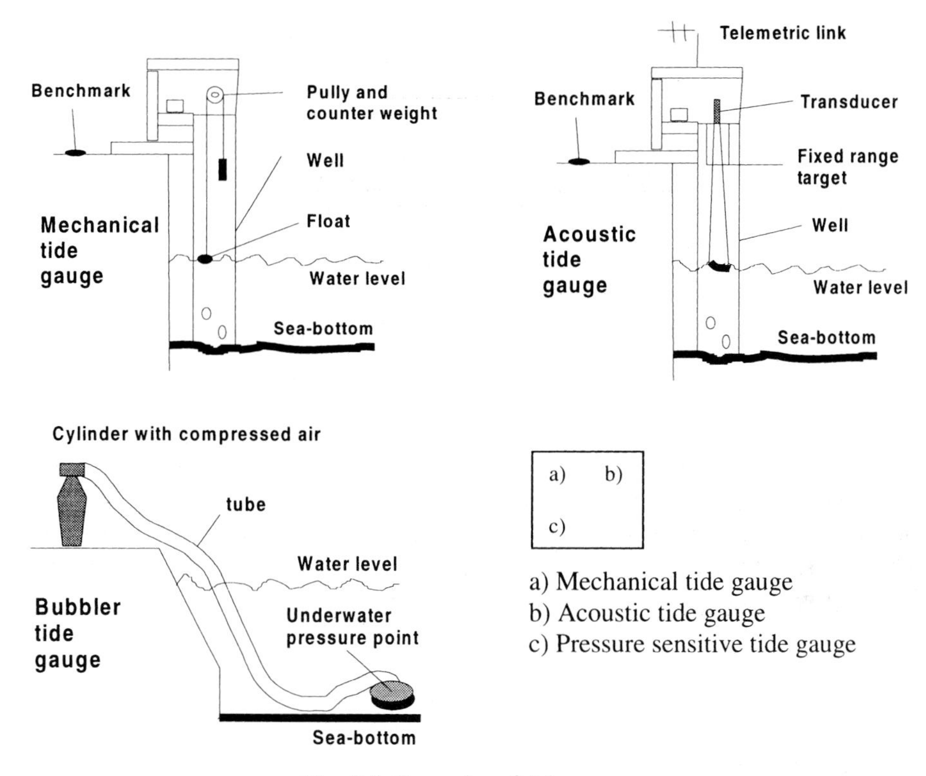

a) Mechanical tide gauge
b) Acoustic tide gauge
c) Pressure sensitive tide gauge

Fig. 8.3 Examples of tide gauges

The tide gauges can be used for different purposes. Most permanent installations are of type 1 or 2 shown in Figure 8.3a and b. In the first case, the tidal observations can be recorded in analogue form on graph paper or in digital form. In the second case, the recorded data can be transferred immediately with a telemetric link from the remote site of the tide gauge to a navigational centre (see Figure 8.3b). This centre collects the information from all available tide gauges. Then the actual height of the water surface above the chart datum can be made available in real-time for the navigation of vessels with radio communication via UHF or telephone line.

Mechanical tide gauges with a float mechanism (Figure 8.3a) are robust and relatively simple to operate. Acoustic gauges (Figure 8.3b) for range measurements down to the water surface have been developed to replace the float type gauges and are very robust and highly accurate. For both types of gauge, measurements are taken in a well or plastic tube to reduce the influence of weather conditions. The well is connected to the water with openings placed below the lowest likely level of the tide. Both of them, however, have similar disadvantages. The installation requires a vertical structure and is rather expensive. Fundamental problems are caused by the physical behaviour of the wells. During the tidal cycle the water level in the well may differ from the level outside because of differences in water density, e.g. at high tide the density in the well is less than in the open sea with the result that the well level is higher. For the second type, calibration for the range measurement is required to compensate for temperature variations. The most effective compensation is to use a fixed range target to compare the measured range with a known distance. Overall an accuracy of ±1.5 cm for the tide record can be achieved.

Pressure operated tide gauges can be used for permanent or remote installations. The bubbler tide gauge shown in Figure 8.3c is portable and fairly quickly installed at low water. It is possible to have the underwater pressure point some hundreds of metres from the recorder on land. These types have a high sensibility due to water density variations caused by changes of temperature and salinity. High accuracy for the determination of the tide variations cannot be achieved near an estuary or river where the water density variations are high. Under good conditions the accuracy of the tide record is a few centimetres in water depths of 10 m. After the establishment of the gauge a calibration against a tide pole is required. There are also other types of pressure operated tide gauges, e.g. seabed pressure gauges for measurements in water depths up to 6000 m.

A basic requirement for all methods is careful siting. The place of tidal observations must be in or very close to the survey area, it should face the open water and not be impounded behind a sand bar, and it must not dry out at low water. In the case where tidal effects are complex or for a large survey area, more than one site is required and needs to be monitored. A connection to the vertical datum has to be established. At best the zero mark of the gauge should be set at the chart datum. The tide record is the basis for the reduction of soundings in hydrographic surveys. There are many different mathematical methods for the estimation of the tidal correction for the actual vessel position. In general, the correction has to be obtained with interpolation from the available data. The accuracy of the tide reduction is given in Table 8.4 (see section 8.3.2.4).

8.2.4 Other Water Level Changes

Apart from tidal variations, water level changes in oceans and estuaries depend on several meteorological influences. Vertical motions are caused by waves, seiches, wind-driven piling up of water, storm surges, swell, and atmospheric pressure changes. Some of them are periodical in some sense, e.g. swell is a long period wave motion created by distant storms. The periods of swell are typically 5–10 seconds and can achieve ranges of up to several metres. The changes vary both spatially and temporally throughout a survey area and can therefore not be modelled.

Wind and weather induce vertical motions (heave) and changes in orientation of the vessel. Depending on the type of survey and the chosen depth measurement system (vertical beam echo sounder or multibeam sonar systems, see section 8.3) the changes have to be taken into account. To reduce their effect on depth measurements, surveys have to be carried out in calm conditions or the errors have to be compensated. Compensation can be achieved when additional sensors for measuring the heave and the vessel's orientation are included in the system. The heave is measured with a so-called heave compensator. The vessel's orientation is described by three angles, i.e., the heading, pitch, and roll. The heading is the orientation of the vessel's fore-aft axis with respect to north and is usually measured with a magnetic or gyro compass. Pitch and roll are two angular displacements of the vessel from level. Pitch is the rotation about the transverse axis of the vessel which results when the nose of the boat goes up or down, and roll is the rotation about the longitudinal axis resulting in sideways waving of the boat. They can be measured using a so-called motion sensor.

For surveys on rivers and lakes the water level is monitored on gauges. River level changes can be as large as (or even larger than) tidal variations. The water level depends on variations in precipitation and runoff where seasonal periods are perceptible between dry and wet seasons. On rivers and lakes dammed for hydroelectric power generation, water levels will vary according to the flow rates through the dam or generator. In wide rivers and large lakes the water level may also be influenced by some of the meteorological effects listed above [Wells, 1997].

8.3 Soundings

Soundings can be defined as the measurements of the depth of a column of water from the surface of a moving platform to the bottom. After Wells [1996 and 1997], they can be obtained as measurements of

- point depths that are isolated depth values on some regular grid pattern or irregular spaced pattern (Figure 8.4a),
- depth profiles along specific sounding lines (Figure 8.4b) or
- area coverage (Figure 8.4c).

The basic idea for achieving area coverage is similar to that in aerial photogrammetry (see Chapter 9). In the case of hydrographic surveying,

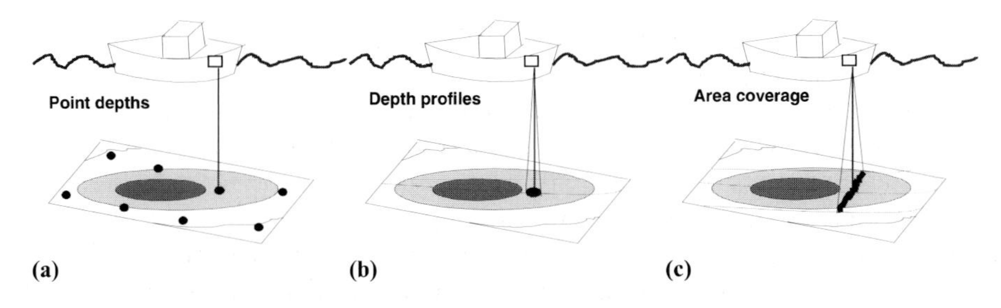

Fig. 8.4 Measuring depths

a survey vessel runs along parallel sounding lines and the sensors obtain a cross-track profile at each sampling epoch (section 8.3.3). The ideal of 100% or complete topographic coverage can only be achieved if the successive profiles are overlapping along track and crosswise. Systems for depth profiles obtain topographic information that is dense along the line, but between the lines no topographic information is available. Interpolation between the lines is necessary to create a Digital Terrain Model (DTM). If isolated depth values are obtained, no topographic information is available between the points.

Starting with the use of sounding rods or lead lines, the principles of modern systems for depth measurements will be described in section 8.3.1. Most of the underwater sensors use acoustic signals. Therefore, the physical behaviour of acoustic waves in seawater will be discussed in more detail. For the measurement of depth profiles vertical beam echo sounders are most commonly employed, and for area coverage multi-transducer sweep or multibeam systems are used. Other techniques, such as sidescan sonar and airborne laser systems, will be briefly introduced in section 8.3.4.

8.3.1 Sounding Rods and Lead Lines

Sounding rods or lead lines can be used for point depth measurements or depth profiles (see Figures 8.4a or 8.4b). They were the only means of making depth measurements before echo sounders emerged, and they are still used in shallow water under the following conditions:

- When echo sounding equipment is not available or cannot fit into the boat which is used for sounding.
- For surveys on small rivers or channels.
- As an aid in the classification of possible false echoes in shallow water and to check the last depth obtained over a pinnacle rock, coral head, or shoal. They are also useful in areas where weeds are growing, or where an extremely soft bottom makes it difficult to determine where liquid water ends and solid bottom starts.
- For obtaining bottom samples where more sophisticated methods are not practicable.

Sounding rods are similar to levelling staffs and are used for depths not greater than 4 m and a water current velocity less than 3 m s^{-1}. The reading of the water level on the staff corresponds to the depth value. The rod is usually made from bamboo, steel, or aluminium and should be strong enough to withstand pressure of the current. The lower end of the rod can be fitted with a metal shoe of sufficient weight to hold it upright in the water and with a base area large enough to prevent it from sinking into the mud or sand.

The use of sounding rods is the most accurate method for sounding in shallow water. They provide direct mechanical measurements of depth, free from the acoustic propagation uncertainties of echo sounders. For applications along a wharf or pier and soundings in a river, distance lines or taut wires can be used to form depth profiles and measurements are taken at regular intervals (e.g. every 5 m, 10 m, etc.). There are some disadvantages, however, such as the measurements being time consuming and the seabed topography being very sparsely sampled as sounding rods provide only one point depth per lowering.

The lead line is a line made of hemp, twine, or thin wire attached to a heavy weight of 5 to 10 kg (usually a lump of lead – hence the name). It is the oldest device for depth measurement and has been in use for over 400 years. To obtain the depth value the line can be marked at regular intervals (0.1 m, 0.5 m, 1.0 m, etc.) or the length of the line can be determined by running it through a measuring gauge as it is paid out.

The lead lines are seldom used for depths greater than 40 m. They also provide direct mechanical measurements of depths. However, there are similar disadvantages as with the use of sounding rods and, in particular, it is hard to detect when the weight has hit the seabed and water currents can carry the line away from the vertical, producing an error in the depth measurement [Wells, 1996].

8.3.2 Vertical Beam Echo Sounders

8.3.2.1 Principle of Operation

The basic operating principle of echo sounders is that an acoustic signal (sound) produced near the water surface will travel to the bottom and be reflected to the surface as an echo (see Figure 8.5). In general, the sound waves are reflected at changes of the density of the carrying medium, e.g. the sea-bottom or riverbed. Echo sounders employ electro-mechanical means to produce a sound signal or pulse, and to receive and amplify the echo. The device for transmission and reception of the signal is called a transducer.

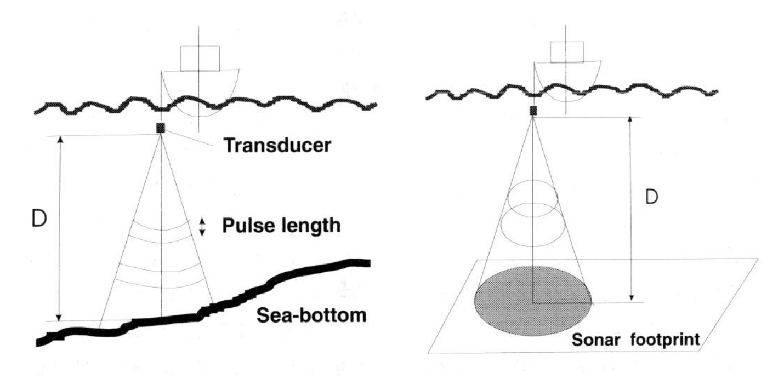

Fig. 8.5 Principle of vertical beam echo sounders

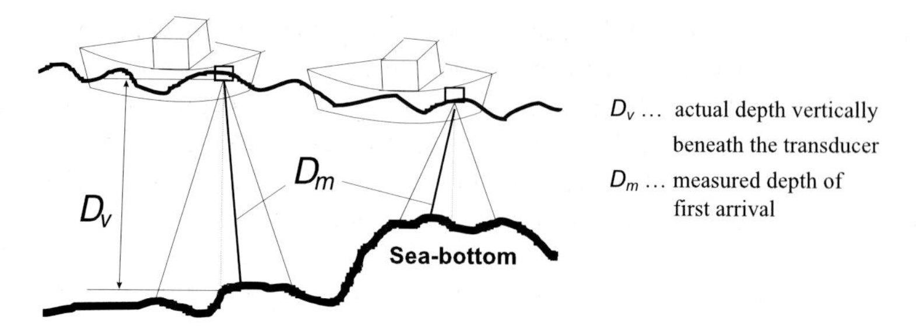

Fig. 8.6 Depth record over rough topography

A transmitting transducer is used to convert the electric power from a pulse generator to an acoustic signal and to transmit the acoustic sound waves into the water medium. The reflected or backscattered sound waves are received at the same or a second transducer and then converted to electric power. The Two Way Travel Time (TWTT) of the signal pulse between transmitting and receiving time is measured and can be converted to units of depth by multiplying with the average sound velocity in water c_W. The measured depth D_m in metres is given by equation (8.1):

$$D_m = c_W \frac{t_L}{2} \tag{8.1}$$

where the t_L is the TWTT in seconds and the average sound velocity in water c_W is given in metres per second. The depths may be recorded continuously either in analogue form on graph paper (see Figure 8.9) or in digital form.

The recorded depth information is always the depth with the shortest TWTT, which is not necessarily vertically beneath the transducer. The recorded signal is called first arrival (or measured depth D_m) and it may differ from the depth vertically beneath the transducer D_v because of slope inclination and roughness of the topography of the sea-bottom (see Figure 8.6). Echo sounders always record the shoalest (shallowest) point within the beam.

8.3.2.2 Characteristics of Acoustic Signal Propagation

Acoustic signal propagation has different characteristics than the propagation of electromagnetic waves. The penetration power, resolution, and accuracy of the depth measurements made with echo sounders are affected by many parameters. The main factors are the speed of sound propagation in the water column, the frequency of the signal, the shape of the echo beam described by the beam width, and the length of the acoustic pulse. In the following these parameters are described briefly.

Unlike surveying on land, electromagnetic waves cannot be used for water depth measurement. Electromagnetic waves propagate best in a vacuum but less well as the density of the medium increases. The higher the electromagnetic conductivity the shorter the propagation range of electromagnetic waves. Acoustic waves propagate not at all in a vacuum but best through solids and liquids. The pulse of energy transmitted into the water is in the form of a sound or compression wave (longitudinal P-wave). Transverse or shear (S-) waves are not transmitted in fluids [Wells, 1996]. The pulse may vary in its frequency, duration, and shape. The quality and suitability of an echo sounder will depend very largely upon how these variables are combined. The repetition rate of the pulse may also be varied to suit a particular requirement.

Depending on the frequency of the sound waves the signal is reflected at different medium densities. Low frequency waves of 5–15 kHz, which are almost in the audible band, will transmit energy over long distances. Their power will not be reduced by solid objects in the water, such as suspended sand, and they are only reflected at higher differences in density. They penetrate quite deeply into the sea-bottom, producing echoes of the underlying layers. A disadvantage, however, is that the transducers for producing these signals are very bulky. Frequencies in the range of 10–30 kHz can detect the medium change of water or silt (liquid mud) to rock very precisely. Echo sounders with low frequencies of 5–15 kHz are employed for surveys in the deep ocean (depths in excess of 15 km). For depths between 1 and 8 km frequencies in the band of 20 to 40 kHz are typically used.

High frequency waves (100–300 kHz) are normally reflected from surfaces with small density differences, such as water to silt. The main advantage is that the pulse can be produced by light and compact electronic equipment. Their power, however, will be relatively quickly reduced in water and they will not penetrate deeply into the sea. They are employed for measurements of depths of less than 1 km. Modern dual frequency (high and low frequencies) echo sounders combine both advantages and are therefore able to detect different layers of density. They can also be used for sedimentation research [Hydrographer of the Navy, 1965–1970].

The size of the ensonified area of the sea-bottom or riverbed, the so-called sonar footprint of the signal (see Figure 8.5), depends on the water depth and the beam width. Transducers are designed to produce beam of sound power of various shapes. Figure 8.7 shows the shape of a typical echo beam. It is very similar to that produced by the headlamp of a car. A reflector is used to concentrate the beam in one direction, although apart from the main lobe there are two side lobes. The side lobes serve no useful purpose and can at times produce false echoes. However, the greatest part of the available power will be concentrated in the main lobe. The beam width is characterized by the beam angle β of the main lobe and depends on the size or diameter of the transducer d and the frequency f of the signal. The beam width β in degrees can be estimated using the following relationship:

$$\beta \approx 65^\circ \frac{\lambda}{d} = 65^\circ \frac{c_W}{f * d} \tag{8.2}$$

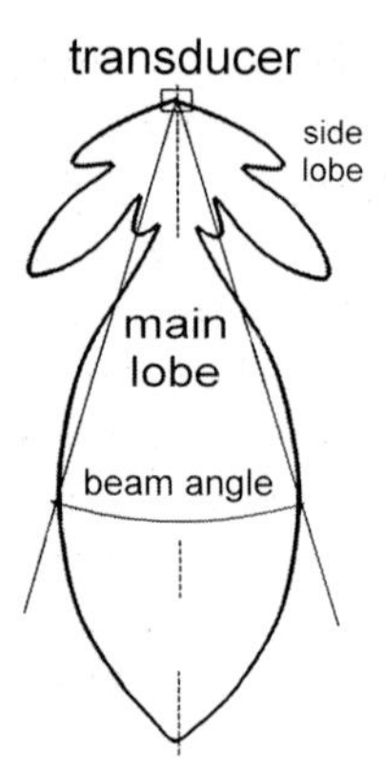

Fig. 8.7 Shape of the echo beam

where λ is the wave length and d the diameter of the transducer in metres or f the frequency in Hertz, and c_W the average sound velocity in water in metres per second [m s^{-1}] [Ingham and Abbott, 1992].

As can be seen from equation (8.2), the higher the frequency the shorter the wavelength and the narrower the beam width for a given size of transducer. A high resolution for the depth measurement goes with the reduction of the size of the sonar footprint and requires therefore a narrow beam width. Large diameter transducers with a high frequency pulse will generate narrow beams at high concentration.

Table 8.2 gives examples of the specifications for the beam width, size of sonar footprint, and diameter of transducer for different kinds of echo sounders. Echo sounders with large beam width β are referred to as wide beam echo sounders and with small β as narrow beam echo sounders. For the calculation of the transducer diameter using equation (8.2) an average value for the sound velocity of 1500 m s^{-1} was used and an acoustic transmission frequency of 30 kHz. The footprint size, however, depends only on the depth and the beam width. The diameter of the footprint is the depth multiplied by the beam angle in radians.

As can be seen from Table 8.2 a narrow beam echo sounder with a frequency of 30 kHz would have a very large diameter and would be very heavy, bulky and expensive. On the other hand, a transducer with a beam width of 30° has poor lateral resolution. Features on the sea-bottom smaller than the footprint size cannot be detected. In practice, for applications in

Table 8.2 Classification of echo sounders

Echo sounder	Beam width	Transducer diameter	Footprint diameter in depth of	
			75 m	1000 m
Wide beam	30°–60°	≈110–50 mm	≈40–80 m	≈525–1050 m
Narrow beam	2°–8°	≈1620–400 mm	≈2.5–10.5 m	≈35–140 m

shallow water a compromise has to be accepted between resolution and beam width. This is the main reason why relatively few narrow beam echo sounders are in use. The resolution also depends on the nature and material of the sea-bottom as they affect the strength of the reflected signal.

The vertical resolution is half the pulse length and is therefore affected by the pulse duration. Two objects in the path of the acoustic ray (e.g. shoals of fishes) that are less than a half pulse length apart would not be detected as two separate echoes. In other words, two separate objects are detected if they are farther apart than half the pulse length. Usually the pulse duration is 1 ms and the wavelength gets 0.1 m using a sound velocity of 1500 m s^{-1}. The pulse length of a 15 kHz signal would be 1.5 m and, in this case, the vertical resolution is then 0.75 m.

8.3.2.3 Estimation of the Speed of Sound in Water

In the previous considerations, an average value for the velocity of sound in water c_W has been used. Following equation (8.1), the sound velocity along the signal path has to be known very precisely to obtain a high accuracy for the depth measurement. The speed of sound in water is a function of its density and elasticity. The elasticity and density both depend, in different ways, on the pressure (or depth), the temperature of the water, and its chemical consistency, i.e. mainly on its salinity. Elasticity usually has a greater effect than density. The greater the elasticity, the faster the speed and, the greater the density, the slower the speed. The elasticity and the speed of sound increase with temperatures up to 40°C. Elasticity will also be increased because of increasing salinity and pressure. On the other hand, water has a maximum density at about 0–4°C. At this temperature the speed of sound is at a minimum and will increase again as the density increases on either side of the density maximum. In addition, increasing salinity and pressure will both increase density and reduce the speed of sound [Hydrographer of the Navy, 1965–1970].

The speed of sound through a water column shows the following sound velocity profile (Figure 8.8, dependence on temperature and pressure at constant salinity only): Near the water surface the speed of sound will be quite high as the temperature is relatively high and the pressure relatively low. As the water deepens, the temperature will generally drop and the pressure will increase. This will reduce the speed of sound with depth and a minimum will be reached at about the depth at which the temperature becomes stable. Below this level, increasing pressure will cause the speed of sound to increase again with depth. Salinity changes will complicate this pattern of change.

Before and after a hydrographic survey the average sound velocity in different water layers has to be determined. This can be done either by direct measurement with a so-called velocity meter, which can be lowered to the sea-bottom to measure the sound velocity at different depths, or by an indirect method, where the temperature, salinity, and pressure are measured

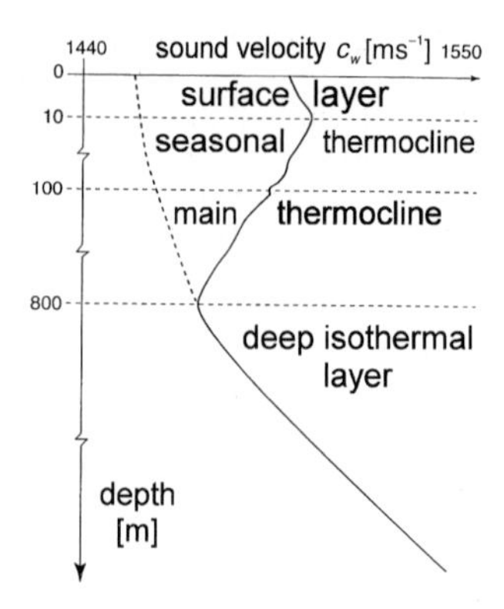

Fig. 8.8 Sound velocity profile

and the velocity is calculated. Using Wilson's equation [Jordon et al., 1966] we get for the sound velocity c_W in metres per second from temperature and salinity measurements:

$$c_W = 1445.4 + 4.62T_W - 0.045T_W^2 + (1.32 - 0.007T_W) \cdot (S_W - 35) \quad (8.3)$$

where T_W is the water temperature in degrees Celsius and S_W the salinity in parts per thousand.

An error in the estimation of the speed of sound causes a systematic error in the depth measurement. The errors accumulate linearly with the water depth. Direct measurement of the sound speed with a velocity meter would provide better results than the indirect method. A velocity meter is very expensive, however, and the measurement of the sound velocity is very time consuming. But also, the indirect method is not very practicable.

In practice, a calibration technique, which is referred to as a bar check [Ingham and Abbott, 1992], is used to obtain an average sound velocity profile. The basic principle is an adjustment of the internally set sound velocity of the echo sounder to the actual sound velocity in the water column, based on a comparison of a precisely known distance with measurements in certain depth sections. First, a bar (hence the name) or disc is lowered beneath a transducer at the maximum depth. Then the measured depth of the echo sounder is compared with the known depth of the bar beneath the transducer and the speed of sound on the echo sounder is adjusted. Afterwards the depth is measured at certain depths at regular intervals (e.g. 1 m, 2 m, 5 m, etc.) to obtain a sound velocity profile and to check the calibration. Normally, the bar check can be used only for calibration in shallow water.

8.3.2.4 Accuracy and Error Budget

As discussed in section 8.3.2.3, the accuracy of the depth measurements depends mainly on the estimation of the sound velocity. The accuracy is

Table. 8.3 Overview of Instrumental Biases

Source of Error	Uncertainty in depth of 5 m	Remarks	Method of Correction
Undetected change of water temperature	±3 cm	$dT_W = \pm 1°C$ $dc_W = \pm 4\,m\,s^{-1}$	calibration
Undetected change of salinity	±1 cm	$dS_W = \pm 1\%$ $dc_W = \pm 1.2\,m\,s^{-1}$	calibration
TWTT measurement itself	±5 cm	resolution of time measurement	–
Beam width of the transducer	dependent on slope angle	–	use of appropriate echo sounder; calculation
Frequency	dependent on quality of sediment	–	use of appropriate echo sounder

Table 8.4 Overview of External Biases

Source of Error	Uncertainty in depth of 5 m	Remarks	Method of Correction
Reduction of soundings due to tidal variations	±5 to ±20 cm	project specific	establishment of inter-mediate tide gauges; mathematical modeling
Change of immersion depth (draught) of vessel or transducer	up to ±5 cm	for a 7 m long vessel, e.g. caused by fuel consumption	adjustment through vessel construction; experimental value
Change of vessel attitude and orientation depending on speed	up to ±5 cm	especially at start and end point of sounding line	adjustment through vessel construction; experimental value
Waves	up to ±10 cm	especially for small vessels	heave compensator
Slope inclination of sea-bottom	±5 cm	for 11° beam angle and 10° inclination	mathematical modeling

degraded by systematic and random errors caused by instrumental and environmental conditions. Table 8.3 and Table 8.4 give an overview of both the instrumental and external biases, their effect on the accuracy of the depth measurement at an average water depth of 5 m, and the method of correction for the survey in coastal areas [Retscher and Teferle, 1998].

Additional error sources for hydrographic surveys in coastal areas caused by the environment are:

- high water pollution due to higher concentration of floating particles,
- high air cavitation and
- undetected and rapid changes in the water temperature or salinity.

Keeping in mind the total error budget an overall accuracy of ±10 to 30 cm for the depth measurement can be achieved with echo sounders in coastal areas. One of the main error sources is the estimation of the current level of the water surface from tidal observations (see Table 8.4). The tide reduction has to be estimated using interpolation for the actual vessel position (section 8.2.2). An error in the estimation of the height of the water level above the chart datum directly affects the accuracy of the reduced depth D_r.

8.3.2.5 Instrumentation

Echo sounders are classified as either wide beam or narrow beam echo sounders (Table 8.2, section 8.3.2.2). Wide beam echo sounders were developed during the First World War and were first used in the 1920s to produce nautical charts. They are still the most widely used hydrographic surveying systems. In an effort to improve the lateral resolution narrow beam echo sounders were developed and they have been available since the late 1950s [Wells, 1997]. A wide variety of transducers and frequencies are

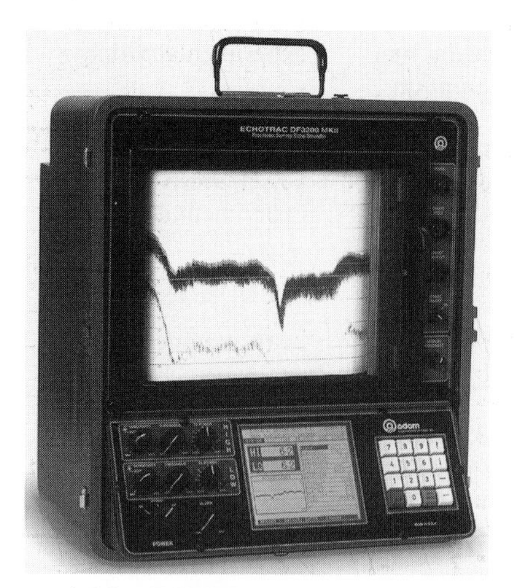

a) Odom ECHOTRAC (Courtesy Odom, Hydrographic Systems, U.S.A.)

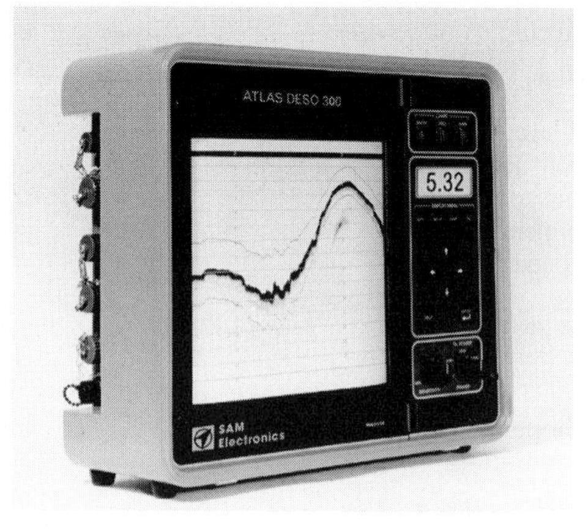

b) Atlas Deso 300 Echosounder (Courtesy STN Atlas Elektronik, Germany)

Fig. 8.9 Examples of vertical beam echo sounders

available for different kinds of applications. Many systems come in both single and dual frequency models. Figure 8.9 shows examples of modern vertical beam echo sounders.

Conventionally a depth profile is recorded from the echo sounder on graph paper. The analogue display chart is referred to as the echo trace. Time or event marks are used to enable the synchronization of depth measurements with positioning of the vessel. Nowadays, digitizers convert the analogue trace information into digital depth soundings.

8.3.3 Multibeam Sonar and Sweep Systems

Starting with vertical beam echo sounders, systems for obtaining area coverage have been developed (see Figure 8.4c). One method is to arrange several vertical beam echo sounders along booms suspended athwartships on either side of a vessel. These systems are called multi-transducer sweep systems (Figure 8.10) and have been in use since the mid 1960s. The most common systems have up to 36 transducers. A cross-track profile is obtained at each sampling epoch and 100% coverage can be achieved if successive profiles overlap. It is essential that the spacing between the transducers and the individual beam width should be matched to the depths being measured, because the size of the footprint should be large enough so that the footprints overlap each other.

The application of these systems is limited to shallow water (normally depths less than 10 m) and they are used for surveys in rivers and harbours. The accuracy of the depth measurement is similar to that of vertical beam echo sounders. The suspended booms, however, require a very slow speed of the vessel and calm weather conditions.

The most advanced systems for area coverage are so-called multibeam sonar systems (see Figure 8.11). Originally they were developed in the mid 1970s for surveys in deep water. Since the mid 1980s systems for shallow water (depths up to 500 m) have also been available on the market.

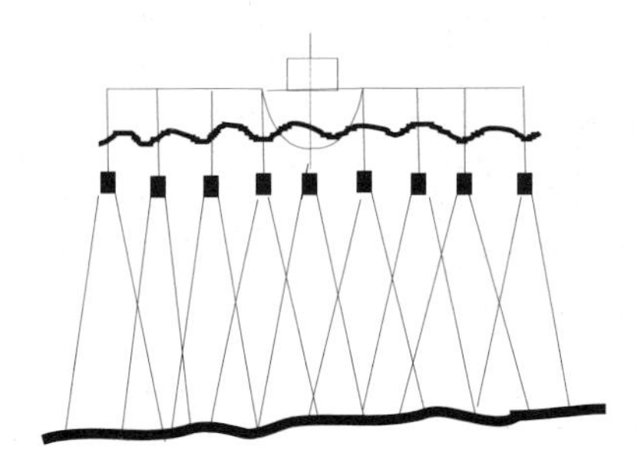

Fig. 8.10 Multi-transducer sweep system

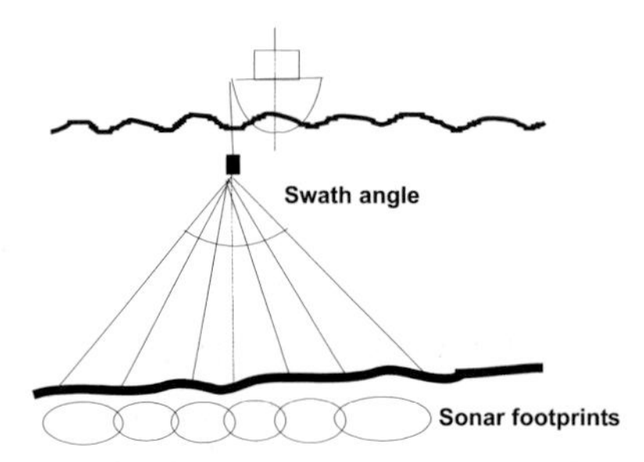

Fig. 8.11 Multibeam sonar system

Multibeam sonar systems consist of an array of equally spaced transducers that produce a fan of narrow beams in a sector, the so-called swath. The fan of sound pulses is transmitted athwartships and it is received in segments on its return. The depths in a swath from each beam can be calculated from the TWTT measurement and the elevation angle of the beam. One hundred percent bottom coverage can be achieved with appropriately designed line spacing and overlap between cross track profiles. The swath width varies as a function of the system type and water depth. Typically swath angles of 90° to 150° are used, which result in a swath width of twice the water depth for a 90° fan and 7.5 times for a 150° fan. Similar to the classification of echo sounders, the systems are referred to as narrow swath and wide swath multibeam systems.

The accuracy of the depth measurement, however, degrades with the increase of the swath angle. Oblique beams with a small elevation angle at the outer side of the fan produce larger sonar footprints, as can be seen in Figure 8.11, and they have a poorer resolution. Therefore, it is often not possible to meet depth accuracy standards out of the full swath width of wide swath multibeam systems.

The application of multibeam systems in deep water is much easier than in shallow water because of their sensitivity to environmental conditions. Especially in the upper layers of the water column the speed of sound changes rapidly and causes depth errors. Oblique beams are affected more by acoustic variations than vertical beams. In addition, vessel motions (heave, pitch, and roll, see section 8.2.4) greatly influence the measurements. To make the multibeam data useful, a motion sensor has to be included with the system to provide compensation for the motions of the vessel. As a result, shallow water multibeam systems are very complex and expensive. Powerful software packages are required to process the massive amount of data that are collected during the survey [Wells, 1996 and 1997].

Figure 8.12 shows typical survey vessels equipped with shallow water multibeam sonar systems and the processing unit on board. The systems use different technologies for the generation of the beam and different shapes for the transducer array (flat or curved arrays). Frequencies in the range of 100 to 300 kHz are typically used.

Fig. 8.12 Shallow water multibeam sonar systems on survey vessels

8.3.4 Other Techniques

Apart from the techniques described above, a variety of other techniques for underwater data acquisition are available. Some of them can be used instead of or complementary with echo sounders for special applications. Other systems are employed for the detection of depth anomalies on the seafloor, e.g. the existence of a navigational hazard or a ship wreck, and for the inspection of the nature of the seabed material, the presence of sand waves and rock outcrops, and geological and geophysical investigations.

8.3.4.1 Sidescan Sonar Systems

The basic construction of a sidescan sonar system is similar to an echo sounder with its transducers tilted obliquely. Then the depth measurements made by an echo sounder become slant ranges. An array of transducers produces a rectangular beam that is thin in the along-track direction and broad in the across-track direction (beam width is usually 40°). As the main lobe is not pointing vertically downwards, the seabed vertically beneath the transducers is usually recorded by echoes from the side lobe (see Figure 8.7). Acoustic backscatter instead of the TWTT is recorded as a function of the elapsed time. The result of the measurement is an acoustic image of the seafloor relative to the position of the sonar. The image, or so-called sonogram, is similar to the negative of an oblique aerial photograph. The transducers are normally mounted on either side of a towfish that is towed behind the vessel at a small height above the sea-bottom (Figure 8.13) [Wells, 1997].

Although the resolution in range is comparatively poor, the sidescan sonar systems have a good detection capability for objects on the seafloor as

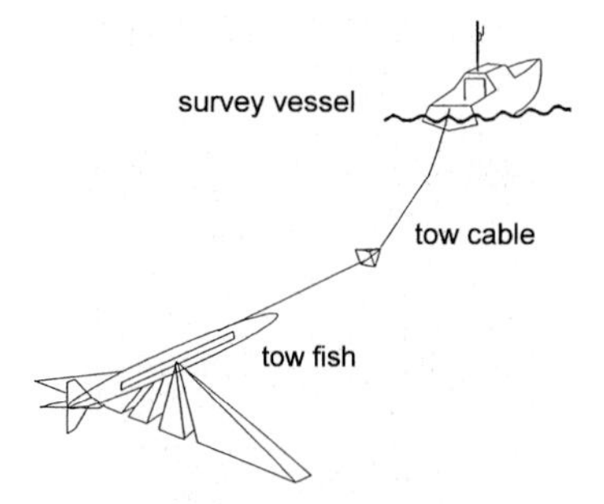

Fig. 8.13 Sidescan sonar system

small as 1 m in dimension using heights of 50 to 300 m of the towfish above the sea-bottom. The nature of the seabed material, rock outcrops, and wrecks can be identified. Applications of sidescan sonar systems vary from geological research in deep oceans to detection of depth anomalies in navigational channels. In coastal areas they can be used to supplement vertical beam echo sounders to check the seabed topography between the sounding lines or depth profiles [Ingham and Abbott, 1992].

8.3.4.2 Airborne Laser Systems

Airborne laser systems can be employed for rapid and high density sounding of shallow water. The systems are referred as LIDAR which stands for **L**aser **I**nterferometric **D**epth **a**nd **R**anging. The systems obtain point depths over a regular spaced grid. Their basic principle of operation is that a laser transmits short laser pulses at two different frequencies (a green and infrared laser pulse) downward from an aircraft or helicopter to measure the water depth. The infrared beam is reflected from the water surface, the green beam penetrates the water surface and is reflected from the sea-bottom. The water depth is determined from the TWTT difference between surface and bottom echoes of the infrared and green signals.

Figure 8.14 shows an example of an operable LIDAR system. The laser pulses scan the sea-bottom in a grid pattern (e.g. $5\,\text{m} \times 5\,\text{m}$) across-track to the flight direction. Swath widths are typically 100 to 250 m using flying heights of 360 to 500 m. Area coverage can be achieved with properly designed flying heights and line spacing. The area coverage rate is typically 20 km^2 per hour at 5 m spot distance. The range of the depth measurement is limited and depends very much on the water quality. In clear water, depths up to 40 m can be easily measured. The accuracy of the depth measurement is about ± 20 to 30 cm. Applications of the airborne laser systems are typically in remote areas or in areas of difficult access for ships, such as archipelagos with numerous small islands, reefs, and narrow ship routes [Steinvall et al., 1997; Wells, 1997].

Fig. 8.14 Laser Airborne Depth Sounder LADS MkII (Courtesy LADS Corporation, Australia)

8.4 Survey Preparation and Data Processing

In the previous sections the fundamentals and operational steps for hydrograpic surveys have been discussed separately. Using modern hydrographic software packages the three main tasks of hydrograpic surveys can be performed and controlled by an experienced operator. The three main tasks shown in Figure 8.15 are:

- the survey preparation,
- the performance of the survey itself, and
- the data processing.

The first part is the survey preparation. This includes the survey design and planning, which will be discussed in the section 8.4.1. Also the field calibration belongs to the survey preparation. The calibration of the survey equipment has to be carried out before the start of the survey. Therefore corrections for the survey devices can be calculated, e.g. the velocity sound corrections for the echo sounder as described in section 8.3.2.3.

After completing the survey preparation the survey work can begin. The hydrographic software package has to collect the raw measurement data from the positioning devices and the depth measurements from the echo sounder. If other sensors are employed, e.g. a motion sensor or a gyro, they have to be included in the measurement process. It is also very important to monitor the route of the vessel. Several devices are available to perform track guidance, e.g. a left/right indicator on a helmsman display (see section 8.4.2).

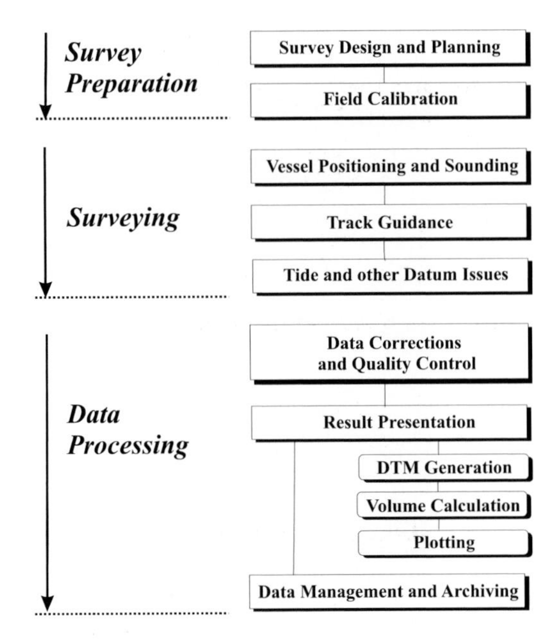

Fig. 8.15 Operational chart of hydrographic surveys

At the end of the survey, the data processing and result presentation is performed. Most of the work is usually done in the office after the fieldwork has been carried out. For hydrographic surveys, however, some of these steps might also be done during the survey. The data processing will be briefly described in section 8.4.3.

8.4.1 Planning a Hydrographic Survey

As for many tasks in surveying, careful planning of hydrographic surveys is an essential requirement. Starting from an analysis of the client's requirements, all available documents of the survey area have to be investigated. For most applications a field reconnaissance is unavoidable and a useful tool for preparation of the sounding operations. Apart from planning the position determination of the hydrographic survey vessel and tidal observations on tide gauges, the main aim of planning is to organize the sounding operations.

The planned lines of soundings (depth profiles) must cover the seabed in the survey area in a methodical manner. Figure 8.16 shows a typical example of planned sounding lines for a coastal survey using a vertical beam echo sounder. For soundings near the coastline small boats are used and for deeper areas survey vessels are required. The sounding lines usually should run at right angles to the coastline or depth contour lines. The lines should be straight and evenly spaced (5 to 10 mm in the plan depending on its scale). The spacing may be varied according to the coverage of the echo sounder

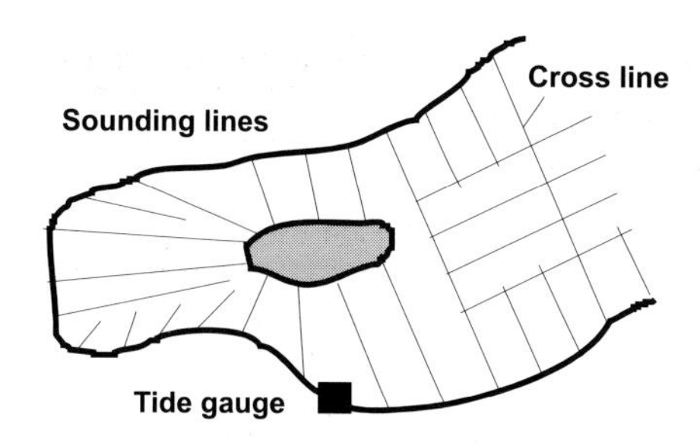

Fig. 8.16 Planned sounding lines of a coastal survey

beam footprint. Additional cross lines at right angles to the sounding lines should be carried out for checking purpose [Ingham and Abbott, 1992].

8.4.2 Track Guidance

Many software packages for on-line navigation purposes are commercially available these days. Figure 8.17 shows a typical navigational display with track guidance features. All essential parameters to navigate the boat along a sounding line are displayed when in on-line mode. These include system time, available hard disk space, communication ports set up, and computer status. Start of Line and End of Line coordinates are displayed with bearing and length of track. In a graphical display the position of the boat is shown in relation to the line and values like cross track error (CTE), distance on line (i.e., 'Länge' in Figure 8.17), distance to go (i.e., 'Zum Ende'), and course made good (i.e., 'Kurs') can be viewed numerically. All specific sounding line parameters, like Start of Line and End of Line coordinates, can be activated in the profile overview window. The guidance for the coxswain of the vessel can be performed using a simple left/right indicator on a helmsman display, where the perpendicular distance of the vessel position from the planned sounding line is displayed. If the boat is left of the planned survey line, the value is negative; if to the right, it is positive [Retscher and Teferle, 1998].

8.4.3 Data Processing

The hydrographic software package has to collect the raw measurement data from the positioning devices and the depth measurements from the echo sounder. If DGPS is used for positioning (see section 7.3), datum transformations from the geographic position in the WGS84 coordinate frame to other local survey datums, e.g. the local grid system, also have to be performed. In addition, tide gauge measurements have to be carried out during the survey. Tide corrections can be applied manually after the survey. Some software packages are able to apply real-time tide corrections from

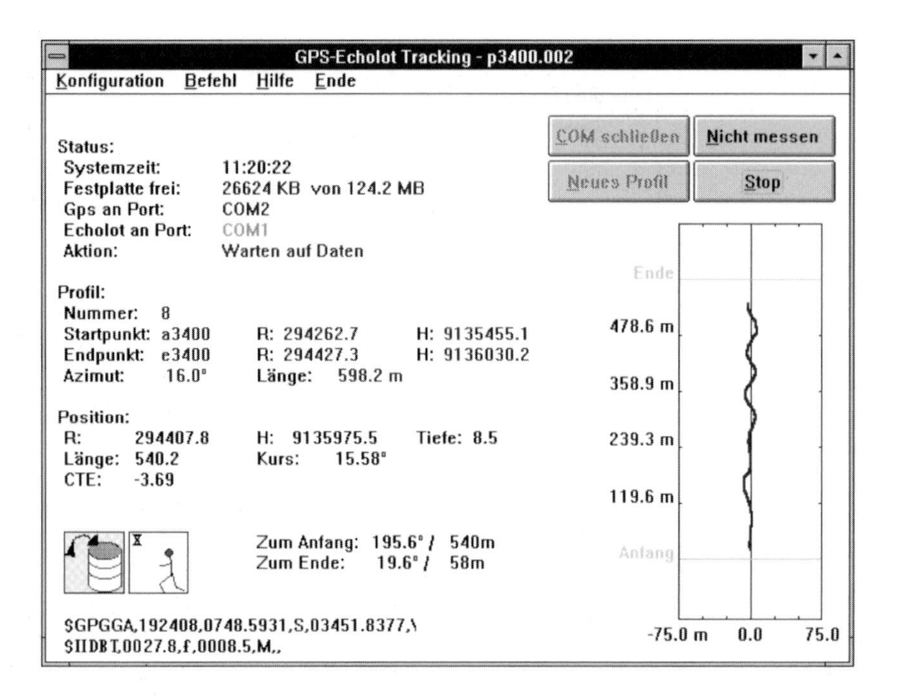

Fig. 8.17 Navigational display with track guidance features

certain tide gauges. These systems send the current water level information to the survey boat via a radio telemetry link at user defined intervals (see section 8.2.3). The software has to read these updates and save them to the raw data file. Then these corrections will be automatically applied during the processing step.

The data processing step includes the reduction of the soundings and data corrections, e.g. manual tide corrections or corrections for sound velocity. Also the quality of the raw data has to be checked. The survey results can be presented as Digital Terrain Models (DTM) or using other visualization techniques. Sometimes volumes also have to be calculated and a nautical chart has to be plotted. The manner in which the results are presented depends of course on the client's specifications.

The last step in a hydrographic survey project might be the data management and archiving. All the data has to be stored for future use (see Figure 8.15).

8.5 Summary and Outlook on Future Developments

Accurate and complete recording of the shape and nature of the sea-bottom is the central theme of hydrography. Although there are a variety of methods that can be used for different applications, usually acoustic echo sounders are employed. Thus the traditional single vertical wide-beam echo sounder remains the most commonly used hydrographic tool for many applications. Because of technological progress in the past few years, multibeam sonar systems were developed to explore the nature of the deep-ocean seabed. These systems are designed to provide 100% area coverage, for the first time. Nowadays these systems are slowly being introduced for surveys in shallow waters with depths up to 500 m. In addition, other techniques, such as sidescan sonars, are employed to inspect the nature of the seabed material, and for geological and geophysical investigations. Laser bathymetry from aircraft may replace conventional techniques in shallow water applications with difficult access for ships, such as archipelagos with numerous small islands, reefs, and narrow ship routes.

Similar to the evolution of echo sounders for underwater data acquisition, DGPS (see section 7.3) has revolutionized the positioning of hydrographic survey vessels and made possible for the first time the real-time track guidance on preplanned sounding lines. Future developments in this area involve the introduction of electronic charts (Electronic Chart Display Information Systems ECDIS) and real-time navigation similar to car navigation systems. They can be employed not only for the guidance of hydrographic survey vessels but also for any positioning demand in the marine community [Wells, 1997]. All these new technologies together will enhance the main goal of underwater data acquisition, the increase of navigation safety in harbour areas and important ship routes all over the world.

9 Image Acquisition

Bruce King and Kent Lam

9.1 Introduction

Images of one form or another provide a significant source of information for GIS. As the development of multimedia technology matures, the integration of images into a GIS environment has become more important. Images can be classified in many different ways, such as by their type (digital or analogue), by the imaging device used to capture them (frame camera or linear scanner), or by the imaging platform that carried the device (aircraft or spacecraft). Section 9.2 discusses the two types of images-digital and analogue. Section 9.3 describes the different types of imaging devices-cameras and scanners. Scanners are divided into two types, those that convert analogue images to digital and those that capture the digital image in real time. Factors specifically relating to air-borne imagery – those taken from aircraft typically with large and medium format cameras (often referred to as aerial photogrammetry) – are addressed in section 9.4, while section 9.5 discusses imagery obtained from space-borne sensors. This imagery typically comes from real-time scanning systems and is known as satellite remote sensing.

It should be noted that this chapter addresses the most commonly used source of imagery for GIS applications. As well as those mentioned here, cameras are used on the ground, in many forms of aircraft (including kites, hot-air balloons, model aircraft, and helicopters), and from manned spacecraft. Real-time scanning systems are also found on and close to the ground, in aircraft as well as on board orbiting spacecraft. There are many sources of images and image types that can find application in GIS. This chapter merely introduces you to the most commonly used ones.

9.2 Image Types

Images may be classified into two types depending on how they were originally acquired:

- Those that were originally recorded onto film – analogue images.
- Those that were originally recorded by an electronic device – digital images.

Apart from this classification, it should be noted that for use in a GIS, it is necessary to have imagery in digital form. This means that those analogue images need to be digitised before they can be used. The conversion is achieved through scanning.

The difference between the image types is the imaging medium. For an analogue image, it is film and for a digital image, it is typically a charge-coupled device (CCD). CCDs are also employed in the devices that are used to convert the analogue image into digital form. In this section the differences between the two imaging mediums are discussed. The properties of analogue sensors (film) will be described and then those of digital sensors.

9.2.1 Analogue Images

Film has been in existence for a long time. The first efforts at capturing an image occurred in the nineteenth century when Louis Daguerre presented his first daguerrotypes – positive images recorded onto metal plates. From then the process of film based imagery or photographs developed. Modern film is far removed from the original daguerrotype, but the fundamental principal remains. It is necessary to identify two different classes of "film" – that which provides a transparent image and that used for making an opaque image. The former is best known as film and the latter as print.

Both images consist of a base material on which an emulsion of silver halide crystals is placed. The difference between the two is this base. For film, it is usually a transparent plastic (polyethylene) material and for print material a polyethylene coated paper. The difference between panchromatic (black and white) film and colour film is the number of layers in the emulsion. Colour photographs are achieved by making different layers sensitive to each of the primary colours.

Silver halide crystals change their state when exposed to light forming a latent image of the scene presented to the film. The greater the intensity of the light the greater the number of individual crystals that are converted. When the film is processed, a chemical reaction results in the converted crystals being removed from the surface of the film. On the film (considering a panchromatic film), the areas of bright light become transparent as the large numbers of converted crystals are removed and the dark areas remain. As the silver crystals are very small a continuous change in the *density* (darkness) of the image is achieved and different shades of grey are produced.

The main characteristic of film is its *sensitivity*. This is determined in two concepts – the film's sensitivity to a portion of the electromagnetic spectrum (spectral sensitivity) and its sensitivity to the quantity of light required for exposure. The first property is used to classify films into broad categories, such as infrared or visible light products, while the second is used to classify the film's response under different lighting conditions. The selection of film with appropriate spectral sensitivity is dependent upon the photographic task and this is beyond the scope of this book.

It is important to select the correct speed rating of the film. That is how quickly a film (or more importantly its emulsion) will respond to a given amount of light. Under low lighting conditions, a *fast* film is required so that a small amount of light is sufficient to expose the film. Under bright illumination a fast film will become over exposed and so a less responsive film is required. How fast a film responds to a given amount of light is specified by its speed – this is given as an ASA/ISO or DIN number. For general photography, film speeds range from slow, 25 ASA, or up to 1600 ASA (fast) and sometimes higher. For a given quantity of light and aperture setting, the use of one film with twice the speed of another film will require half the exposure time. For aerial films a slightly different rating scale, called the Effective Aerial Film Speed (EAFS), is used.

The speed of a film is related to the size of the silver halide crystals. Typically slower films have smaller crystals. A side effect of using slower films is that the resultant images can be enlarged to a greater extent than can images from faster films before the grain of the film becomes apparent.

An issue associated with film speed is the *resolving power* of the film – the ability to accurately image small objects, or closely separated objects. While not the only parameter affecting resolution, film speed and the related grain size play significant roles. This is not a critical issue for general photography, but it is significant for aerial photography. The basis of resolution assessment is the quantity of line pairs per millimetre (lppm). One line pairs per millimetre consists of a pair of black and white lines occupying a space of one millimetre. Aerial photographic systems are capable of achieving over 200 line pairs per millimetre with the final result being a combination of all the elements in the imaging process.

A more meaningful measure of the resolution of an image is the *modulation transfer function* (MTF). The MTF indicates the relationship between lppm and the relative difference between the brightness of the original object and the acquired image.

9.2.2 Digital Images

Whereas film is capable of recording an infinite variation of light intensity, a digital sensor can only do so in discrete steps. This is because of the need to convert a quantity of light into a digital number. An image that would be recorded as a photograph on film is instead recorded as a matrix of digital numbers, each representing the intensity of light at that point in the image. The whole digital image is made up of data obtained from a sensor element.

A sensor element may be simply imagined as a bucket. The role of the bucket is to collect electromagnetic energy (most commonly visible light, but it may be other parts of the electromagnetic spectrum) over a fixed period of time. The amount of energy in the bucket at the end of the time corresponds to the brightness of that part of the scene being imaged by the sensor element. Depending on the design of the sensing device, it may be made up of a single element (bucket), a linear array (line of buckets), or a matrix of elements (grid of buckets). The three types of sensors will be described in detail in section 9.3.2. When displayed, each sensor element is often referred to as a *pixel*, which means picture element.

The output from the sensor is a string of digital numbers in binary form. Each digital number (DN) corresponds to the brightness (intensity of energy) of the sensor element. Most commonly, the digital numbers are eight bits long (2^8), which gives DNs from 0 to 255. For a panchromatic image, 0 correspond with black and 255 with white and the image is said to have one *colour plane*. For a colour image there are three colour planes, one each for red, green, and blue, and the image will have a total of 24 bits which produces 2^{24} (16 777 216) different colours. The conversion from light intensity to digital number is achieved by an analogue to digital (A/D) converter.

The colour depth (bits per pixel, radiometric resolution) depends on the sophistication of the sensor. The number of bits may be interpreted as the sensitivity of the sensor; more bits indicate that the sensor can discriminate smaller differences between levels of intensity of the incident energy. Also, sensors may be tuned to be sensitive to particular portions of the electromagnetic spectrum just as films are.

When the strings of DNs are correctly assembled and displayed on a computer monitor, it is seen as a digital image. The software displaying the image reads the digital number and converts it back into an intensity that is then placed on the monitor in its correct relative position to the other pixels. An example of a digital image and digital numbers is shown in Figure 9.1.

Just as the resolving power of film is related to the grain size of the emulsion, so the resolving power of a digital sensor is related to the size and spacing of the sensor elements. In early digital systems there were gaps between sensor elements, and so information falling on the image plane of such systems was lost. In current systems, there is 100% coverage of the image plane, so it is sensor size that is important. Imaging systems typically produce pixel sizes as small as 5 μm with experimental systems smaller than this. At the pixel level this indicates that, at best, a resolution of only 100 lppm may be achieved and more typically 50–70 lppm is the reality. From this point of view, analogue images offer superior resolution to digital ones of the same scale.

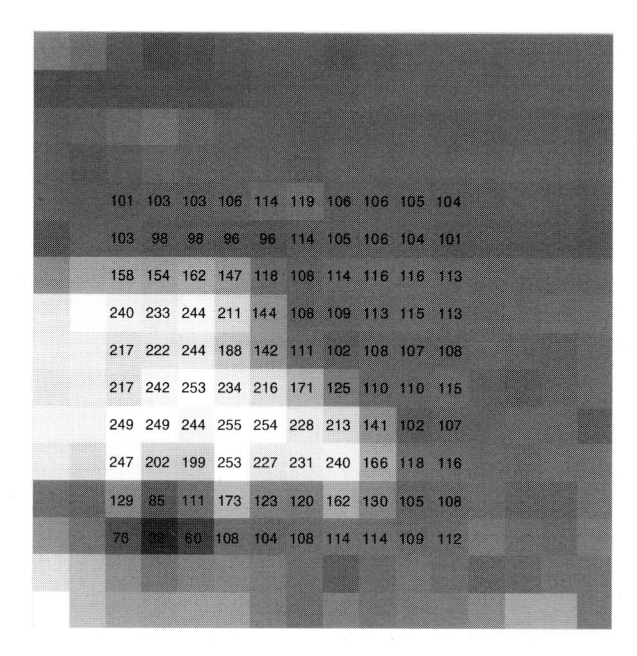

Fig. 9.1 A digital image

9.3 Imaging Devices

Imaging devices take two different forms: cameras and scanners. Cameras may create either analogue or digital images but scanners, as they are used today, generate only digital images. The fundamental difference is the way in which an image is acquired. Cameras capture a whole scene in a single, almost instantaneous exposure. Scanners build up the image pixel by pixel, line by line, or patch by patch. This section looks at the components that make up cameras and scanners.

9.3.1 Cameras

A broad range of cameras may be used for the acquisition of images that are used in GIS. At this time, the most commonly used camera is a *single lens frame camera*, which is a camera that exposes an area of film (the image area) through a single lens. Other types of cameras include multi-lens frame cameras, panoramic cameras, and strip cameras. Digital single lens frame cameras have some application in this area, but cannot yet compete with their film based brothers for efficient, high resolution capture of large areas of terrain.

Much of the analogue imagery used in GIS comes from the aerial mapping camera. This type of camera produces large format photographs that typically have a nominal image area 9 inches (230 mm) square. Because of the image size, these cameras are commonly called *large format* cameras. These cameras are highly sophisticated in their design and construction in order to produce high resolution, low distortion photographs that are used for the creation of mapping products such as topographic maps and orthoimages. Other frame cameras have a smaller image size. Medium format cameras (also often called professional cameras because they are used by professional photographers) have image sizes of the order of 60 mm × 60 mm and 35 mm cameras (small format or amateur cameras) have an image size of 36 mm × 24 mm. Examples of large, medium, and small format cameras are shown in Figures 9.2, 9.3, and 9.4, respectively.

Digital cameras may be classified according to the size of the imaging sensor. The sensors are area arrays ranging in size from typically VGA resolution of 640 × 480 pixels up to high resolution cameras with 4096 × 4096 pixels. Lower end digital cameras typically are of the automatic type, which produce images of acceptable quality, but have limited application for photogrammetry. Higher resolution cameras are based upon medium or small format camera bodies and have all the user-controlled features of the corresponding film cameras.

Fig. 9.2 Zeiss RMK TOP15 large format camera (From Zeiss catalogue)

Fig. 9.3 Rollei Q16 medium format digital camera (Rollei brochure)

Fig. 9.4 Kodak DCS small format digital camera

9.3.1.1 Camera Components

The most sophisticated form of camera (from the point of view of GIS imaging) is the aerial mapping camera. These cameras are highly sophisticated cameras of metric quality. The features of an aerial mapping camera include:

- Large film capacity.
- Precision film flattening by means of a vacuum.
- Fiducial marks to locate the principal point and indicate dimensional stability.
- Auxiliary data blocks recording imaging details.
- Calibrated principal distance.
- Calibrated lens distortion parameters.
- Forward motion compensation (FMC) to eliminate image blur due to camera motion while the shutter is open.
- Camera stabilisation device to keep the camera pointing vertically down.

Medium and small format cameras designed for photogrammetry will have similar features. They may have either fiducial marks or a *reseau plate* – a glass plate placed in the image plane that contains a series of crosses that produce a fine grid on the image. As well as serving a similar function to the fiducial marks mentioned earlier, the function of the grid is to allow the correction of any local distortion that may occur between the time the image is acquired and when it is used. These cameras will not have either FMC or stabilisation devices.

If the location of the fiducial or reseau marks, the principal distance, and the lens distortion parameters are known and are stable, then the camera is called a *metric* camera. If all or some of these parameters are known or considered to be unstable, the camera is called *semi-metric*. If none of this information is known it is called a *non-metric* camera. The parameters are determined through a process called *calibration*.

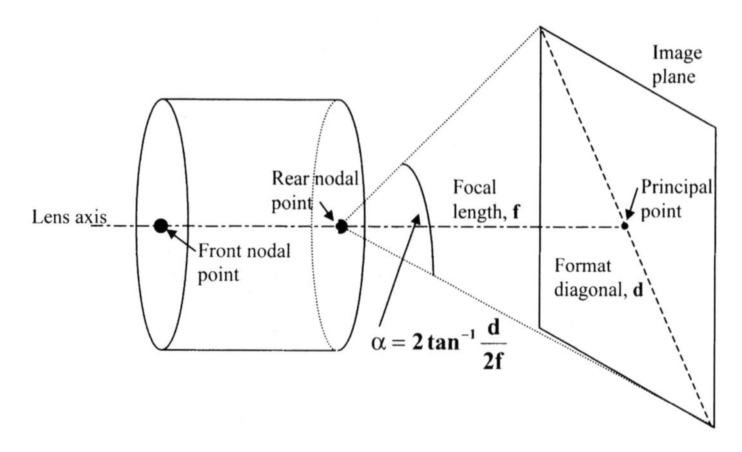

Fig. 9.5 Angular field of view

Table 9.1 Angular field of view for common aerial camera lenses

Focal length	85 mm	152 mm	305 mm	610 mm
Field of view	125°	94°	56°	30°
Class	Super-wide	Wide	Normal	Telephoto

9.3.1.2 Image Classification

The images taken by an aerial camera are typically categorised by its *field of view*. This is the angle subtended at the rear nodal point of the lens by the diagonal of the image format as shown in Figure 9.5. When the principal distance and the image diagonal are similar in size, then this is called a *standard* or *normal* angle lens. For aerial photography the relationship between some lenses, their fields of view, and classification are given in Table 9.1.

9.3.1.3 Lens Quality

The quality of a lens is described by various parameters. Two most widely used parameters – particularly for photogrammetric purposes – are resolution and distortion. As mentioned previously, resolution is measured in line-pairs-per-millimetre or MTF. For a given lens it should be noted that resolution varies across the field of view and is typically highest at the axis of the lens.

Distortion consists of two components – radial and tangential. When looking at the image plane, radial distortion displaces the imaged point away from its undistorted location along a line radial from the principal point, while tangential distortion displaces the image point in a direction perpendicular to the radial line containing the undistorted point. These are illustrated in Figure 9.6.

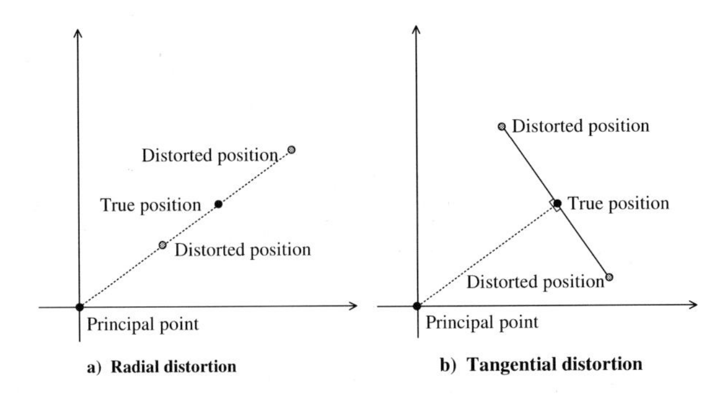

Fig. 9.6 Radial distortion (left) and tangential distortion (right)

Radial distortion is the result of imperfect grinding of the lens elements. It is symmetrical about the principal point and may be as small as 10 μm for a well built aerial camera lens and as much as 300 μm at the edge of the format for a non-metric camera lens. Tangential distortion is attributed to the imperfect alignment of lens elements and is symmetrical about an axis through the principal point. Tangential distortion is not significant for aerial cameras and can be as much as 100 μm for other lenses. In practical terms, the distortion of aerial imagery is negligible and it is only for the most demanding applications of data extraction that it must be accounted for in the data extraction process.

The resolution of a lens and the amount of distortion are determined by calibration. It should be noted that the amount and pattern of distortion varies as the focus of the lens changes. As lenses used for aerial cameras are of fixed focus, then calibration from time to time is sufficient. When using lenses that can be focused, however, it is important that the distortion be determined for the exact focal setting used to acquire the imagery.

9.3.2 Scanners

This class of image acquisition device differs from the camera in that it sequentially obtains the digital image rather than a simultaneous capture. Scanners take several forms including those that are quite familiar to us as desktop and film scanners, and those less familiar such as those used in satellite remote sensing systems. Desktop and film scanners are used to convert existing analogue imagery into digital form. During the process for document scanning, the analogue imagery is placed in a control or semi-control environment and remains static in a fixed position throughout the entire scanning process. For real-time digital image capture in space borne and air borne scanning missions, however, both the scanner head and the scanning platform are in motion. Images captured by these systems are under the influence of the scanner head and platform motions as well as environmental changes.

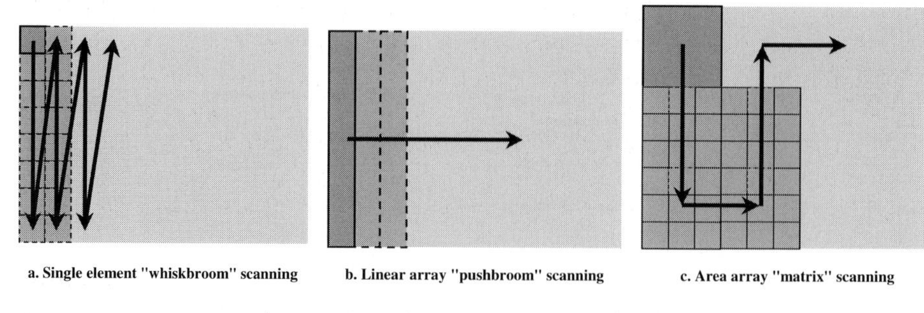

Fig. 9.7 Basic sensor configurations

Regardless of their function, scanners build up an image by moving the sensor past the object. The sensors used in scanners may take three forms:

- A single sensor element that is moved across and along the object – a whiskbroom scanner.
- A linear array that captures a *swath* or strip of the image in one instant – a pushbroom scanner.
- An area array that captures a *patch* of the image at one instant – a matrix scanner.

Each of these three forms is illustrated in Figure 9.7. Depending on the area to be imaged, the linear and area arrays may be moved in one or two directions to capture the entire scene. How the sensor is used depends upon the design of the imaging system.

Each type of sensor has many applications ranging from being incorporated in desktop scanning systems, to digital cameras, to real-time imaging systems such as satellite remote sensing systems. Digital cameras have already been discussed in previous sections; desktop scanning systems and real-time scanning systems are discussed here.

9.3.2.1 Desktop Scanning Systems

For use in GIS, two types of documents are of prime importance. These are:

- photographs
- maps and plans.

Photographs may either be scanned from paper prints, diapositives, or negatives. Maps and plans may be scanned from the original medium (usually a plastic based material) or from paper prints. There is an extreme range to the cost of document scanning systems ranging from low-end desktop scanners designed for general purpose use to high-end scanners specifically designed for scanning photographs, maps, and plans. The prime difference between the two is the geometric accuracy of the scanned image. Scanners designed specifically for photogrammetric and map scanning applications typically have very high accuracy and precision, which is necessary to preserve the exact relationship in the digital version of the photograph that existed in the original. The cost of achieving this is high.

The meaning of accuracy in this case is the difference between the position of a pixel in the resulting scanned image compared to the image point's position in the original photograph. Precision is a measure of how much variation in position of a single pixel might be observed with repeated scanning of one image.

Scanners may be of either flatbed or drum configuration. Flatbed scanners have the limitation of material size, whereas drum scanners can handle very large documents. Another significant difference between the two is how the scanning is done. Flatbed scanners typically hold the document stationary and move the sensor, whilst drum scanners move the document over the sensor. For large drum scanners, both the sensor and the document may move.

Low-end film scanners and drum scanners are designed to convert small and medium format film images into digital images. These scanners are designed primarily for digital artists and so, also, do not focus on geometric accuracy and precision. Another significant difference is the size of the scanned area. Desktop scanners can handle paper usually of US Letter size or 8.5″ × 11″ low end film scanners 35 mm, 120 roll film and 4″ × 5″. Photogrammetric scanners need to accept the full 230 mm size of an aerial photograph. Photogrammetric scanners are built for the large rolls of film that are used in aerial cameras. They will automatically scan images and advance the roll. A comparison between typical desktop and photogrammetric image scanners is given in Table 9.2.

The accuracy and precision of photogrammetric film scanners is ensured by rigorous design and construction methods and the use of quality components. As the cost of scanning equipment decreases, then so does the accuracy and precision. However, low-end commercial scanning systems

Table 9.2 Comparison of commercial scanners

Scanner	Type	Cost (US $)	Document size	Optical image resolution	Image precision	Image accuracy
Hewlett-Packard ScanJet 5100C	Document-flatbed	200	216 mm× 297 mm	300 dpi, 85 μm	Not stated	Not stated
Intergraph PhotoScan TD	Photo-grammetric	170,000	250 mm× 250 mm	3,600 dpi, 7 μm	1 μm	2 μm per axis
Minolta QuickScan 35	Film	1,000	24 mm× 36 mm	2,820 dpi, 9 μm	Not stated	Not stated

need not be rejected for applications requiring moderate levels of geometric accuracy. In this case, the scanner may be calibrated to determine the systematic geometric errors of the device and image processing techniques such as re-sampling (see Chapter 12) can be applied to the image to correct for these errors. As these scanning systems are usually installed in either a semi-controlled environment for low-end commercial systems or a fully controlled environment for high-end drum and film scanners, the scanning process is not affected by changes in the environment, such as humidity, temperature, and lighting effects. Besides, as document or film mounted on the scanning surface remains unchanged throughout the scanning process, fewer systematic errors are introduced into the imagery. On the contrary, real-time scanning systems are operated in a dynamic environment where many systematic errors are imbedded into the imagery during the scanning process. Due to the instability of the scanning platform in air borne scanning systems, images captured by these systems suffer from a larger scale of systematic errors than images captured from the more stable space borne scanning systems.

9.3.2.2 Real-Time Scanning Systems

Real-time scanning systems usually operate from two different platforms: air borne and space borne. For air borne scanning systems, the scanner is mounted in an aircraft. The scanning process takes place while the aircraft is flying in the air at a constant velocity. As the aircraft motion is affected by wind direction and magnitude, the platform may become unstable which will affect the quality of the captured images. On the contrary, for space borne satellite based scanning systems, as the scanner is mounted on a satellite that flies in a stable and precisely pre-determined orbit, fewer errors are introduced in the captured images.

There are two different ways in which remote sensing scanners (sensors) can build up two-dimensional images of the terrain – using across-track (whiskbroom) scanning or along-track (pushbroom) scanning. Both sensors are line scanning devices with the whiskbroom scanner sensing one pixel at a time using a single detector but the pushbroom scanner sensing a linear array of pixels at a time using an array of detectors. The two best known satellite systems, Landsat and SPOT, employ these two different sensing systems to scan the terrain.

The whiskbroom scanning system is an optical mechanical system, which scans the terrain along lines perpendicular to the orbital track with the use of an oscillating mirror. The sweeping motion of the scanner is synchronised with the forward motion of the satellite to yield a series of contiguous strips that are combined to produce a two-dimensional image. A single detector is used to capture the incident energy focused on it. At any instant during the scanning process, the detector focuses on a small area of the terrain. Energy emitted or reflected from that small area is detected by the scanner. An on-board A-to-D signal converter is used to convert and record the average

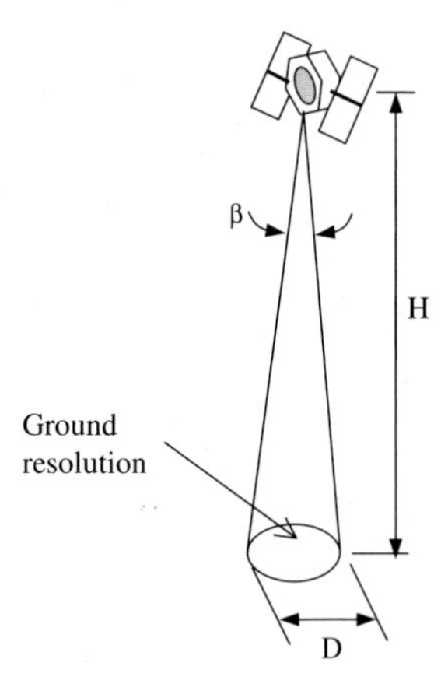

Fig. 9.8 Instantaneous field of view and the resultant ground resolution

energy in digital number (usually ranging from 0–255). The size of the focused area is controlled by the system's instantaneous field of view (IFOV) expressed as a cone angle (β) in proportion to the flying height (H) of the spacecraft. Figure 9.8 illustrates the geometric relationship of the three elements. The size of the area can be expressed as a circle of diameter (D) given by:

$$D = H\beta \tag{9.1}$$

where D is the diameter of the circular ground area viewed; H is the flying height above the terrain; β is the IFOV of the system (expressed in radians). The diameter D of the ground area is commonly referred to as ground resolution and is generally approximated by the side of a square (pixel resolution).

The coverage of an image is governed by the pixel resolution, the total field of view, the flying height, and the number of scans. The total field of view is defined as the total angular motion allowed in one swath. The Landsat multispectral scanner (MSS) is a typical example of an across-track scanning device. It scans the terrain only during the forward sweeping motion. Figure 9.9 illustrates the MSS operating configuration. A nominal scene consists of 2340 scan lines with about 3240 pixels per line. A full scene of MSS image with four channels consists of 7 581 600 pixels per channel or over 30 million observations. To improve the system response time, six contiguous lines are scanned simultaneously with each mirror oscillation. A 4×6 array of detectors consists of the four spectral bands with six lines for each band being used.

The Thematic Mapper (TM) carried by Landsat-4 and 5 is an advanced multispectral scanning sensor with higher spatial resolution, improved geometric fidelity, and radiometric accuracy than the MSS. The TM is intended to replace the ageing MSS. Its oscillating mirror scans in both the forward and reverse sweeping motions. A scan line corrector located in front

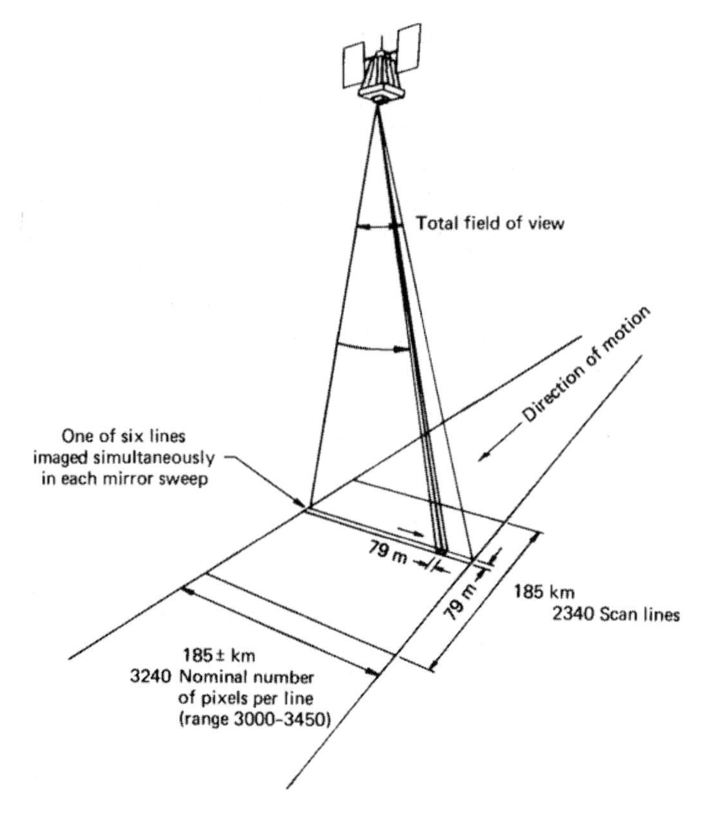

Fig. 9.9 Lansat MSS – an example of cross-track scanning system (from Lillesand and Kiefer [1994])

of the focal plane is there to compensate the forward motion of the spacecraft in order to ensure that scan lines are perpendicular to the orbit.

For the along-track scanning systems, a one-dimensional linear array of detectors (charge coupled devices) are located in the focal plane of the sensors' optical system. The entire array of data is obtained simultaneously during scanning by sampling the incident energy focused on the focal plane and is converted into an array of digital numbers. A series of contiguous scan lines is obtained as the satellite moves over the earth. The pushbroom scanning has certain advantages over the optical mechanical whisk-broom scanning. As there are fewer moving parts in the scanning system, the pushbroom scanners are expected to have a longer life expectancy and use less power to operate. With the pre-calibrated geometric relationship among pixels in the linear array, which is identical for all scan lines in an image, the geometric fidelity of the image is higher. However, the disadvantage is that each detector in the array needs to be individually calibrated to produce a spectrally uniform image scene. For example, the pushbroom scanner in the SPOT satellite has an array of 6000 detectors to be sampled in the panchromatic mode and 3000 detectors in the multispectral mode.

9.4 Air Borne Imagery

9.4.1 Image Characteristics

Although not appreciated by the average photographer, there is a characteristic to the images that allows the extraction of three-dimensional information from them. Whether they are analogue photographs obtained from aeroplanes or digital images obtained from satellites, their common feature is that they are *perspective projections*. That is, associated with each image point is a perspective centre through which the ray of "light" passed on its way from the object to the recording medium (which may be either film or a digital sensor.) The difference between the types of image depends on how much of the image is related to a single perspective centre. From a photogrammetric point of view this is the very property that allows three-dimensional information to be extracted from a two-dimensional image.

It should be noted that images obtained directly from both cameras and scanners are perspective projections. The principles developed here apply equally to all frame and line type images and are not confined to those from aerial imaging devices.

9.4.1.1 The Perspective Projection

The perspective nature of a photograph is illustrated in Figure 9.10. Each point on the three-dimensional surface covered by the photograph is imaged onto the image plane. The significant parameters are the focal length of the imaging system (f), the size of the image, and the distance between the *perspective centre* (the centre of the imaging system's lens) and the object. The focal length is more correctly called the *principal distance* and is defined as the perpendicular distance from the image plane to the rear nodal point of the lens. The point at which this line, known as the *camera axis,* intersects the image plane is called the *principal point*. The principal point can easily be found from a photograph containing *fiducial marks* by intersecting opposite marks as illustrated in Figure 9.11.

The result of a perspective projection is that, for terrain that contains relief, the scale of the image is not constant and that points at the same horizontal position but with different elevations will appear at different locations on the image. These two characteristics are called *scale variation* and *relief displacement*, respectively, and are elaborated on in sections 9.4.1.2 and 9.4.1.3.

Other terms related to an aerial image are *exposure station* and *nadir point*. The exposure station is the point in space occupied by the perspective centre, whilst the nadir point is the point vertically beneath the exposure station. The nadir point can be found on the ground and on the image. For a truly vertical aerial image, it will be the point on the ground that is seen at the principal point.

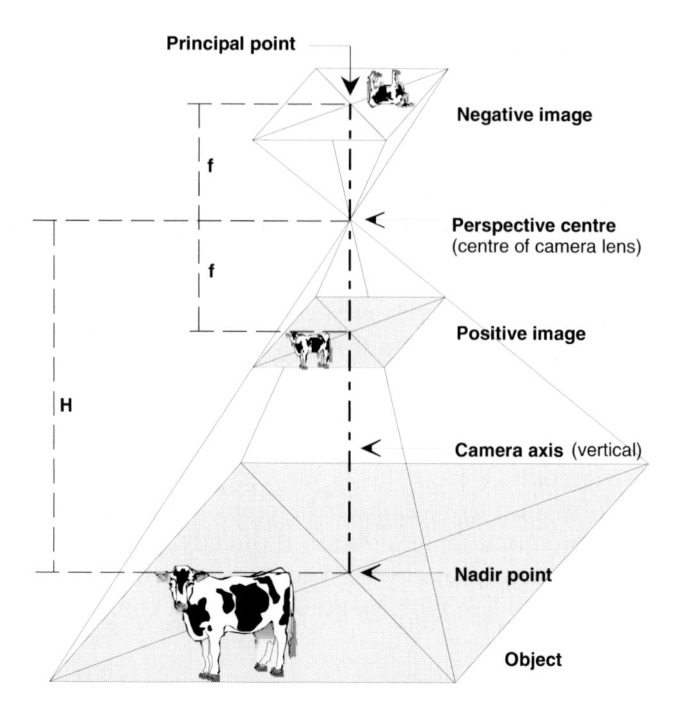

Fig. 9.10 The photograph – a perspective projection

9.4.1.2 Scale of the Photograph

In the case of an aerial photograph such as that obtained for topographic mapping (as shown in Figure 9.11), the axis of the camera is usually directed vertically downwards. As for a map, the scale of a photograph is the ratio of the distance on the photograph to the distance on the object. Unlike a map it can also be defined by the ratio of the focal length of the imaging system to the distance between the exposure station and the object. Thus each point in the photograph has a scale dependent upon its elevation. Hence a photograph of terrain with significant differences in elevation will have a large variation in scale, whereas a photograph of relatively flat terrain will have a relatively uniform scale over the image. Referring to Figure 9.12, the scale of a photograph can be found from:

$$s = \frac{r}{R} \tag{9.2}$$

where r is a distance on the photograph and R is the corresponding distance on the object (ground).
For point A on the ground:

$$s_A = \frac{r_a}{R_A} = \frac{f}{H - h_A}. \tag{9.3}$$

For point A′ on the datum:

$$s_{A'} = \frac{r_{a'}}{R_A} = \frac{f}{H}. \tag{9.4}$$

Similar expressions can be found for points B and B′. So it can be seen that points with the same elevation (such as those on the datum) will have the same scale, whereas those with different elevations will have different scales. This gives rise to the concept of point scale of a photograph – the scale at a particular point – as expressed by equation (9.1). In quoting the scale of a photograph, some sources ignore the height of the terrain and simply use equation (9.3), whereas the more correct quotation of scale will use the approximate average height of the area photographed as shown in equation (9.4). Using (9.3) does not present any difficulties when the height of the terrain is close to that of the mapping datum, but this can give a very misleading result if the imaged terrain is at a height considerably above (or below) the datum.

$$s_{ave} = \frac{f}{H - h_{ave}}. \tag{9.5}$$

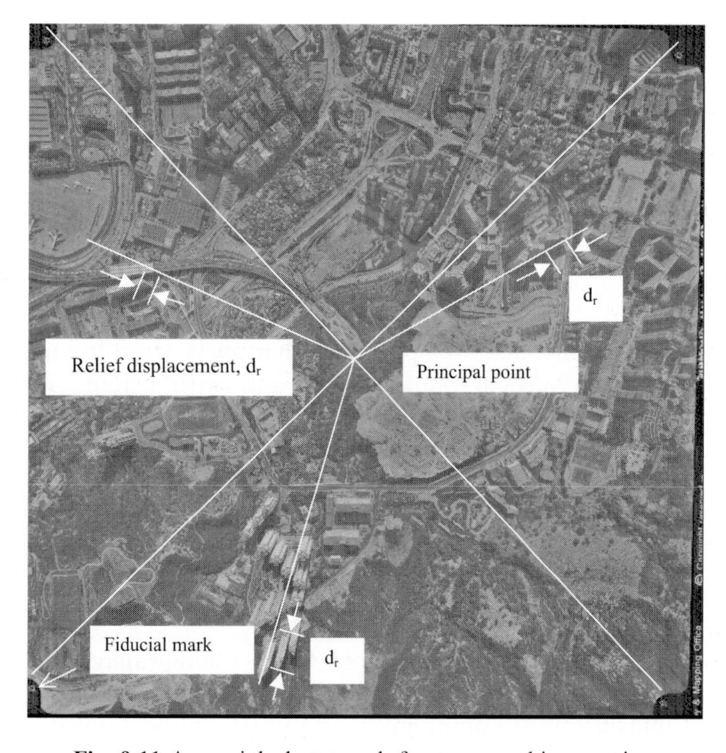

Fig. 9.11 An aerial photograph for topographic mapping

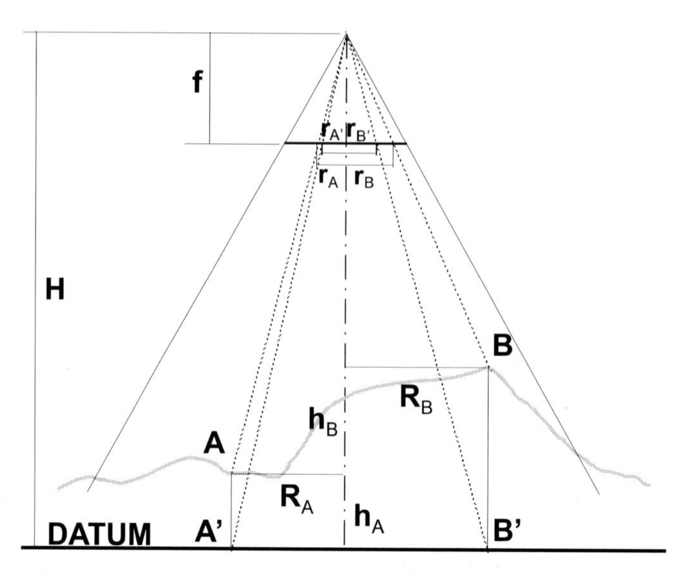

Fig. 9.12 Scale of a photograph

9.4.1.3 Relief Displacement

The fact that scale is not uniform over the whole of a photograph alone is a good enough reason for not treating a photograph as a map or plan. For a photograph that contains only smooth terrain and few or no artificial features, what is not apparent is the more serious consequence of this scale variation-relief displacement. This concept is seen in Figure 9.11 and elaborated on in Figure 9.13. Points at the same horizontal locations but with different heights are imaged at different locations. The equation describing relief displacement is:

$$d_r = \frac{r_b h}{H}. \tag{9.6}$$

Relief displacement increases for points farther away from the nadir point (principal point for vertical photographs) and with height. Taking the photographs from a greater altitude can reduce relief displacement on average across the entire photograph. It must be remembered though that to maintain image scale when using this methodology, a lens of greater focal length must be used. Relief displacement is the characteristic that allows for the extraction of height from overlapping images (Chapter 11).

9.4.2 Strips and Blocks of Aerial Photography

The type of imagery most commonly used in GIS applications comes from aerial photography. The main purpose of this imagery is for topographic

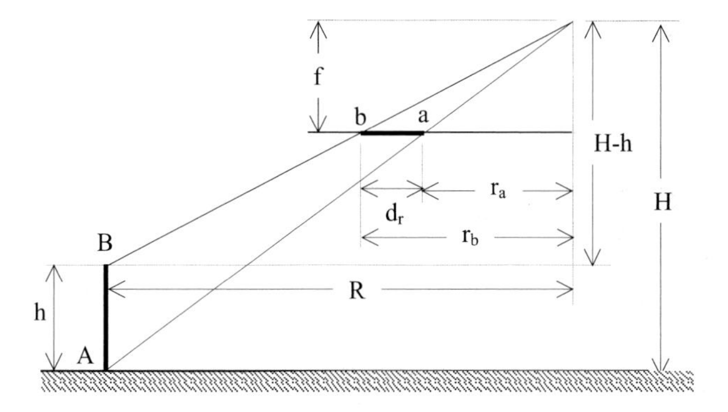

Fig. 9.13 Relief displacement

mapping and interpretation. The basic characteristics of aerial photography are discussed in this section.

The camera is mounted in an aeroplane and large areas of the earth's surface are photographed by flying in a regular pattern of overlapping strips. This is called a block of photographs. The geometry relating to the configuration of strips and blocks is shown in Figure 9.14.

Depending on the average scale of the photograph (s_{ave}) and the size of the image (g), the photograph will have a certain ground coverage, G, given by:

$$G = \frac{g}{s_{ave}}. \tag{9.7}$$

The amount of overlap along the strip is called *endlap*. Endlap is the difference between the ground coverage and the distance the aircraft has moved forward (called the *camera base*, B). It is usually expressed in terms of a percentage and is calculated from:

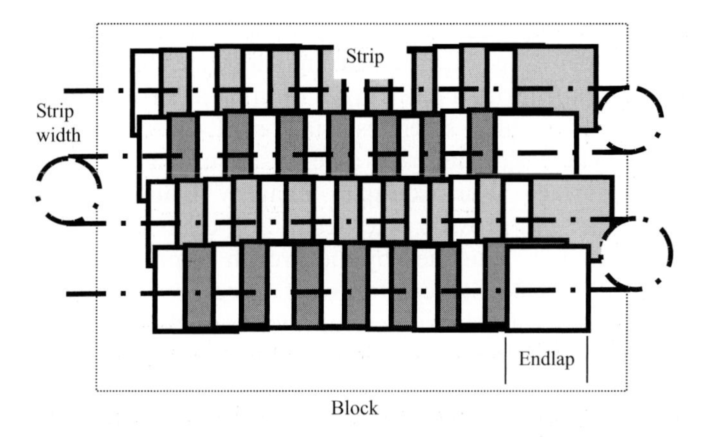

Fig. 9.14 Strips and blocks of aerial photographs

$$E\% = 100 * \frac{(G - B)}{G}. \tag{9.8}$$

To aid stereoscopic viewing, the endlap of aerial photographs is typically 60%.

The distance between adjacent strips that form a block is called the *strip width* (W). The difference between the strip width and the ground coverage is the amount by which adjacent strips overlap and is called *sidelap* (S). Sidelap is computed from:

$$S\% = 100 * \frac{(G - W)}{G}. \tag{9.9}$$

To ensure continuity between strips they will overlap by typically 30%.

Endlap is controlled by the frequency of the shutter operation of the camera and the speed of the aircraft. Sidelap is controlled by the planned spacing of adjacent strips.

9.5 Space Borne Imagery

Space remote sensing began as early as the late 1940s when small format cameras were carried on rockets and satellites. By today's standards, the quality of the captured photographs was very poor to be used for analysis purposes. Images with reasonable quality have been available since the early 1960s when efforts were made at imaging the earth's surface from space using weather satellites of the U.S.A.'s National Oceanic and Atmospheric Administration (NOAA). Coarse views of cloud patterns and the earth's surface were imaged. During that period, images were mainly used by meteorologists for weather forecasting. With the technological advancement in satellite image sensors, the imaging power from space has been significantly improved. Images with medium to high pixel resolution covering a wider range of the electromagnetic spectrum can be acquired and transmitted to the ground receiving station on a periodic basis. Since the early 1970s, earth resource satellites, such as Landsat and SPOT, have been launched. These satellites are able to acquire good quality images of the earth's surface. For example, imagery from Landsat satellites launched by NASA can provide pixel resolution equivalent to 30 m on the ground. SPOT satellites, launched by the French company Centre National d'Etudes Spatiales (CNES), can provide pixel resolution as high as 10 m. Most recently, with the successful launch of the IRS-1D satellite by the Indian Remote Sensing (IRS) program, five-metre pixel resolution can be achieved. Several commercial systems with one metre panchromatic and four metre multispectral spatial resolutions are planned to be launched before the year 2000.

9.5.1 Remote Sensing System Characteristics

Most of the earth resource remote sensing satellite systems (programmes) are specifically designed to acquire data about the earth's resources on a systematic, repetitive, global, and multispectral basis. Images covering the same terrain under the same programme should look very much the same geometrically so that changes on the earth's surface can be identified using image processing techniques discussed in Chapter 12. Change detection is an important function of these satellite programmes for environmental monitoring of the earth's surface. Different orbital parameters, such as flying altitude, orbital angle of inclination, orbital period, and sun synchronisation are set for each programme to ensure the mission can be achieved.

A programme consists of one or more satellites, with each satellite usually carrying more than one sensor system; communication links; attitude-control and orbit-adjustment systems; and a power supply; in communication with a network of ground control stations. The sensor systems may have different characteristics such as spatial and spectral resolution to capture images for different applications. Most of these sensors employ either across-tack (whiskbroom) scanning or along-track (pushbroom) scanning to build up two-dimensional images of the terrain. Captured images are stored onboard for later transmission to the ground control stations. Transmission links can also be used to receive commands from the ground control stations for orbit adjustments and for special remote sensing missions.

9.5.2 Characteristics of Satellite Imagery

As both the across-track and along-track scanning systems are line scanning devices, they share similar geometric characteristics and distortions. Geometric distortions on the scanned image can be introduced by platform perturbations along the orbit as well as by the scanning device. Platform perturbations are created by unwanted motions of the carrier caused by air turbulence, change of wind direction, change of the carrier's velocity, etc., during data collection. These will cause misalignment of the scan lines and create gaps and unwanted overlaps in the image. Fortunately, satellite imagery is almost free from these problems because of the stability of the space orbit. Imagery from airborne sensors is probed to have these problems. Other known distortions, such as tangential scale distortion and one-dimensional relief displacement, are discussed in sections 9.5.2.1 and 9.5.2.2.

9.5.2.1 Tangential Scale Distortion

The tangential scale distortion affects the geometric accuracy of a pixel along a scan line. Scale along a scan line is constantly decreasing towards the edges

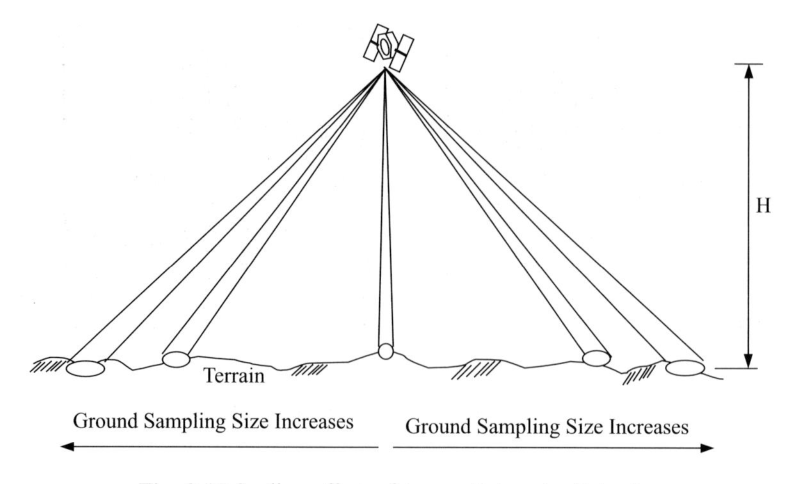

Fig. 9.15 Scaling effect of tangential scale distortion

of a scan. Hence, the ground coverage by a pixel at the edge of a scan line is larger than a pixel near the nadir. The distortion is caused by the increase in the distance between the scanner and the ground. Figure 9.15 illustrates the scaling effect as a result of this distortion. This distortion only occurs along the scanning direction but scale in the direction of the satellite orbit is essentially constant. The one-dimensional progressively compressed effect is illustrated in Figure 9.16. This scale distortion generates a phenomenon called resolution cell size variations. As the scale decreases towards the edge of a scan line, the size of the ground resolution cell near the edge is larger than those near the nadir line. Hence, only larger objects on the ground can completely fill the IFOV near the edge. The same object near the nadir will appear in more than one pixel. However, an advantage of the tangential scale distortion is that the change in ground resolution compensates for the problem of off-nadir radiometric falloff.

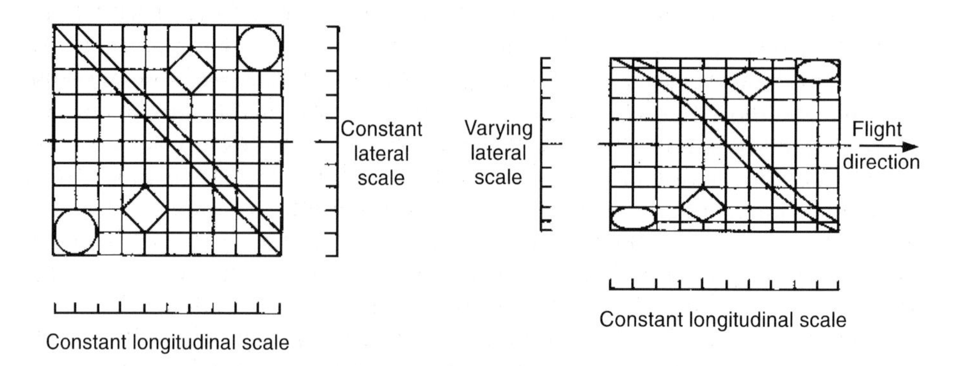

Fig. 9.16 One-dimensional progressively compressed effect (from Lillesand and Kiefer [1994])

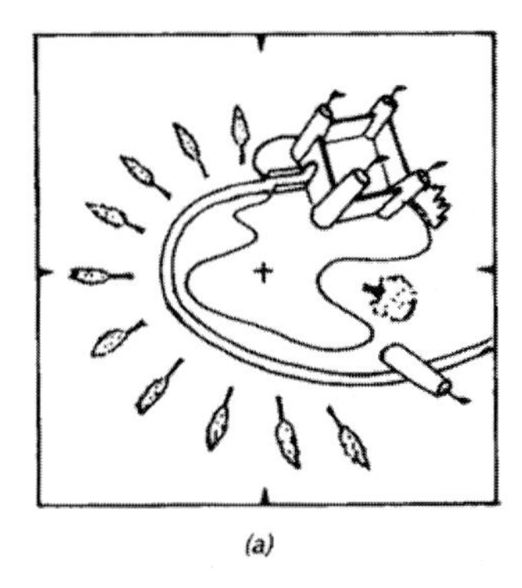
(a)

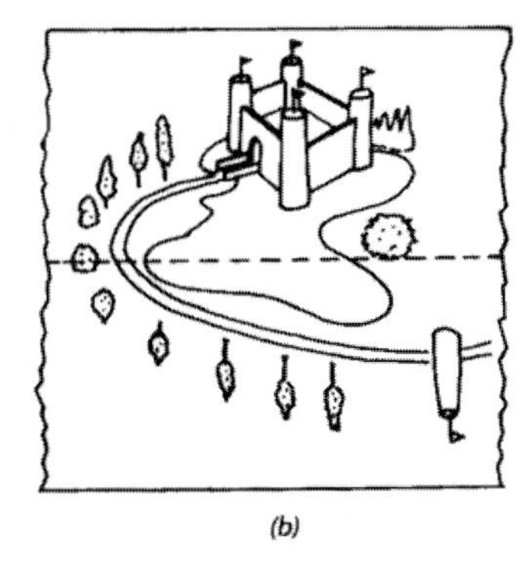
(b)

Fig. 9.17 Relief displacements (from Lillesand and Kiefer [1994])

Tangential scale distortion decreases as the flying height increases and the angular field of view decreases. As for satellite imagery, the extremely high altitude and the very narrow angular field of view produce very little distortion. For example, Landsat satellites have an orbital altitude at about 700 km or higher with the angular field of view no more than 15°. Therefore, for some applications where high geometric accuracy is not required, this type of distortion does not need to be corrected.

9.5.2.2 One-Dimensional Relief Displacement

One-dimensional relief displacement is a very special characteristic of images captured by line scanning devices. Figure 9.17 illustrates the nature of one-dimensional relief displacement in a line scanner image versus relief displacement on a vertical photograph. During the scanning process, objects on the ground are viewed from a "side-looking" direction by the scanner. As a result, relief displacement occurs in the image in only one direction. Therefore, the topes of tall structures along a scan line will appear to be farther away from the nadir line (Figure 9.17b). Similar to the tangential scale distortion, relief displacement decreases as flying height increases. Tall structures will have larger relief displacement than lower structures on the ground (Figure 9.17a). The problem with one-dimensional relief displacement in satellite imagery is not very obvious except for very tall structures on the earth sampled near the edge of an image. Stereo viewing is possible with side-overlapping satellite images, for example, Landsat images captured by adjacent orbits with 14% overlap on the equator and as high as 81% near the two poles. One hundred percent stereo viewing is possible for SPOT images where scanners in the satellites can be programmed for off-nadir viewing. That is, sensors in the SPOT can scan the same terrain fully covered by adjacent satellite tracks.

9.5.2.3 Geometric Accuracy of Satellite Imagery

The spatial resolution of satellite imagery defines the nominal ground coverage sampled by the scanner for a single pixel. This resolution sometimes is referred to as the ground sampled distance (GSD), which defines the smallest object that could be resolved in a digital image. For an object with a width equal to one GSD to be distinguishable from a similar object with the same width, they need to be separated by at least a distance equivalent to one GSD. Therefore, the best ground resolved distance (GRD) would be equal to twice the size of GSD. This is under the assumptions that the image captured is free from optical distortions, extreme illumination, atmospheric effects, and other factors. However, a ground feature smaller than one pixel can also be seen in the image. A single object with extreme contrast from a uniform background may be detected easily but the apparent image size will be distorted. Furthermore, a long narrow object can be identified more easily than a rounded feature. On the contrary, two large objects with less than one pixel spacing may fuse together and become indistinguishable.

A satellite image is subjected to many different distortions, such as atmospheric refraction, relief displacement, earth's curvature, and tangential scale distortion, which affect the accuracy of the image. The geometric correction process is normally implemented in two parts. Systematic distortions that are predictable and can be modelled are corrected first. Then a geometric re-sampling process is applied to correct random and residual systematic distortions, which are unpredictable and cannot be modelled. Well-distributed ground control points (GCPs) are required for this process. A geometrically rectified image will have a higher geometric accuracy than a raw image. Generally, there is a misconception about the use of spatial resolution. A satellite image with spatial resolution of 10 m does not imply that it has a locational accuracy of 10 m. For example, a rectified SPOT image (processing level '2') with spatial resolution of 10 m gives an improved locational accuracy of only 50 m. Satellite images can be purchased from distributors with different data pre-processing levels. For SPOT imagery, some systematic corrections are applied to level '1' data. Ground control points are used for geometric corrections in a level '2' image. Therefore, a level '2' image will have higher locational accuracy than a level '1' image.

10 Orthoimage Generation and Measurement from Single Images

Zhilin Li

10.1 Introduction

Images can be obtained from cameras, scanners, and other types of imaging systems, as discussed in Chapter 9. The most widely used images, however, are photographs taken by air-borne cameras and images taken by space-borne scanners. These images are a major source of data for GIS applications.

In many applications, images are usually used as backdrops, base maps, or as a source from which some types of spatial information can be extracted. To make images useful as backdrops or base maps, it is desirable that the images have characteristics similar to those of maps. Images that do have similar properties are called orthoimages. The generation of an orthoimage, called geo-referencing image in this context, is one topic of this chapter, as is the extraction of useful information from images.

The extraction of information could be through either measurement or interpretation. This chapter discusses the principles and procedures necessary to make images useful as base maps and to make measurements from images. The interpretation of these images, however, will be discussed in Chapter 12. The principles and procedures discussed here are for single images only. Measurements from overlapping images (stereo-pairs) will be discussed in Chapter 11.

Section 10.2 will discuss the principles and procedures for the transformation of single images into an image that can be used as a backdrop a base map (orthoimage). Section 10.3 will describe measurements from single images.

10.2 The Principles for the Generation of Orthoimages

To make images useful as backdrops and/or base maps, it is desirable that the images have characteristics similar to those of maps. That means that the same scaling, north direction, and projection into a geo-referencing system (e.g., a national geodetic system) should be adopted. To accomplish this, a number of requirements must be fulfilled, i.e.,

a) all image points should be registered in a geo-referencing system such as a national geodetic (or grid) system;
b) every point (pixel) of the resultant image should have the same scale if the ground area is small; or else the scale variation should follow a map projection; and
c) the relative relationship between features should also be retained.

If all the images taken by scanners or cameras had such characteristics, then there would be no need for this section. However, as discussed in Chapter 9, images do not have such good characteristics due to distortions caused by the imperfection of camera or scanner systems, the instability of platforms (tilts and flying height variations), atmospheric refraction, earth's curvature, and terrain height variations. The two most serious factors are the instability of the platform and terrain height variations. The process used to correct all these distortions to make the images have characteristics similar to those of maps is called "geo-referencing" in this context. However, this term (geo-referencing) is a little more generic than that currently used in the GIS community. In the current GIS community, geo-referencing refers to a two-dimensional linear transformation to transform two-dimensional image coordinates to a two-dimensional geo-referencing system. In such a case, the image distortion caused by terrain height variation is not considered. This is also called "warping" in some literature, "registration" in the digitisation of images, and simple "rectification" in photogrammetry. On the other hand, the term "geo-referencing images" means a two-dimensional to three-dimensional linear transformation to transform two-dimensional (2-D) image coordinates to a three-dimensional (3-D) geo-referencing system. In this way, the distortions caused by terrain height variations will be considered. In photogrammetric practice, this is also known as "ortho-rectification," "differential rectification," or "DTM-based rectification" since the terrain heights are represented by a digital terrain model (DTM) (see Chapter 11 for a more detailed discussion of DTM generation). The resultant image from such a geo-referencing process is called an "orthoimage" or orthophoto if the images are photographs.

In orthoimage generation, two key steps need to be taken. The first step is to transform every (x,y) coordinate of image points in the image coordinate system into ground coordinates in a geodetic system. This step is called geometric transformation. This is the most important part. However, for practical reasons, there are other processes to follow. All of those processes will be discussed in the following sections to form a complete picture of geo-referencing images.

10.2.1 Geometric Transformation

In the previous section, it was mentioned that the positions of image points are distorted by a number of factors, which will be discussed in section 10.4.

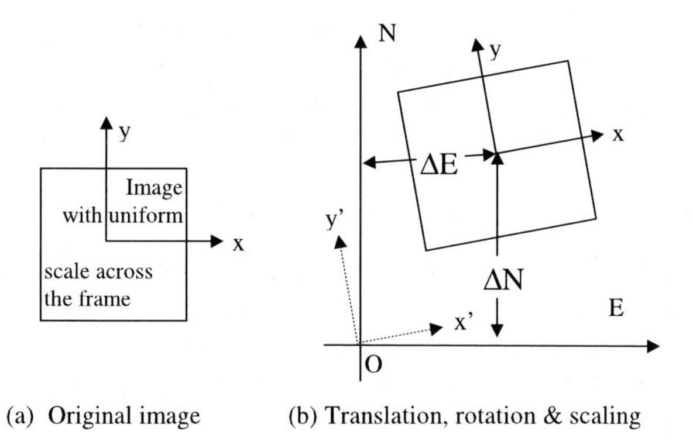

(a) Original image (b) Translation, rotation & scaling

Fig. 10.1 Geo-referencing an image with uniform scale across the frame using a simple 2-D geometric transformation

If such distortions were uniform across the whole image, then there would be a uniform scale change and thus the only problems would be scaling, rotation, and translation, as illustrated in Figure 10.1.

The general mathematical model to transform image points in an image coordinate system to a geo-referencing system is as follows:

$$\begin{aligned} E &= g_x(x, y) \\ N &= g_y(x, y) \end{aligned} \tag{10.1}$$

where, (x,y) are the image coordinates and E (Easting) and N (Northing) are the corresponding coordinates in a geo-referencing system. However, in the rest of this chapter, X and Y will be used instead of E and N for a more general annotation.

However, such an ideal image is difficult to obtain. A geometric transformation that is more complicated than that shown in Figure 10.1 and expressed by equation (10.1) is required to transform image points into a geo-referencing system. There are two ways to deal with the problem. A simple line of thought is (a) to make a correction (Δx and Δy) for each image point and (b) then apply the model expressed by equation (10.1) to transform all (x,y) into (X,Y). The new mathematical model can be rewritten as follows:

$$\begin{aligned} X &= g_x(x + \Delta x, y + \Delta y) \\ Y &= g_y(x + \Delta x, y + \Delta y) \end{aligned} \tag{10.2}$$

where, Δx and Δy represent the corrections for x and y image coordinates caused by all factors.

In the practice of aerial photography, lens distortion for aerial photography is not a problem if professional cameras are used. However, if a low quality camera or scanner is used, lens distortion could be a big issue. In fact, the most serious factors are the tilts of images and the terrain height variations. The corrections for these factors could be modelled in the mathematical model expressing the geometric relationship between 2-D image points (x,y) and the corresponding 3-D ground points (X,Y,Z) (see image formation in Chapter 9 for detail). Such a model would be written as follows:

$$\begin{aligned} X &= g_x(x + \Delta x, y + \Delta y) \\ Y &= g_y(x + \Delta x, y + \Delta y) \\ Z &= g_z(x + \Delta x, y + \Delta y). \end{aligned} \tag{10.3}$$

In equation (10.3), the Δx and Δy have different meanings from those in equation (10.2). Here, the correction for tilts and relief need not to be included.

In the case of orthoimage generation, it is a normal practice that single images (i.e., not a stereo-pair) be used. From a single image, height information cannot be obtained. In this case, the Z value of every point (X,Y) on the ground for equation (10.3) should be known beforehand. Such Z values can be interpolated from an existing DTM or computed from a pair of stereo-images (see a discussion of this in Chapter 11). Therefore, for orthoimage generation, this equation can be rewritten as follows:

$$\begin{aligned} X &= f_x(x + \Delta x, y + \Delta y, Z) \\ Y &= f_y(x + \Delta x, y + \Delta y, Z). \end{aligned} \tag{10.4}$$

Equations (10.3) and (10.4) are general forms of the relationship between image points and their corresponding ground points. In photogrammetric practice, such a model is expressed either by a function called Collinearity Equation (see Chapter 9 or Wolf [1983]) or by a function called DLT (direct linear transformation) [Abdel-Azizi and Karara, 1971]. They are essentially the same, however. They are the different realisations of the projective transformation model. Also, for simplicity, the corrections Δx and Δy will not appear in similar formulae any more because they are so small that they can be neglected for most GIS applications.

In the collinearity model, six orientation elements (three rotation and three translation) are used for an image, either photograph or a scan of image. For photographs, there are six orientation elements to be determined. However, for an image acquired by linear scanners (Chapter 9), each scan line should have six such orientation elements to be determined (because each scan line is acquired at a different time). There is a rotation (φ,ω,κ) around each of the axes of a three-dimensional system (X,Y,Z) and a translation of the origin of the image coordinate system (X_s,Y_s,Z_s).

$$\begin{aligned} X &= f_x(X_S, Y_S, Z_S, \varphi, \omega, \kappa, x, y, Z) \\ Y &= f_y(X_S, Y_S, Z_S, \varphi, \omega, \kappa, x, y, Z). \end{aligned} \tag{10.5}$$

The meaning of equation (10.5) can be understood in this way: two rotations are required to bring an image into its desired position (i.e., a horizontal plane) and another rotation is required to make the x axis of the image coordinate system parallel to the X axis of the ground coordinate system. X_S, Y_S and Z_S are the translations required to make the origins of the two coordinate systems coincide. There are six orientation elements in this model and there is a need of at least three control points for the computation of coefficients.

In this equation, the image coordinate system is as shown in Figure 10.1. That is, the origin being the centre of the image (more precisely the point at which the principal optical axis intersects the image plane) and the x-axis being along the flight direction. Such a definition of an image coordinate system is for the convenience of correcting image distortions. In fact, it is also possible to use an arbitrary Cartesian coordinate system in the image plane. In this case, equation (10.5) can be rearranged as follows:

$$x = \frac{A_1X + A_2Y + A_3Z + A_4}{A_9X + A_{10}Y + A_{11}Z + 1}, \quad y = \frac{A_5X + A_6Y + A_7Z + A_8}{A_9X + A_{10}Y + A_{11}Z + 1} \tag{10.6}$$

where A_1-A_{11} are the coefficients; x and y are the coordinates of an image point; and X, Y and Z are the coordinates of its position on the ground. This formula is called direct linear transformation (DLT) [Abdel-Aziz and Karaka, 1971]. There are 11 coefficients in this model and there is a need of at least 6 control points for the computation of coefficients.

If the terrain is flat, the Z values in equation (10.6) can be set to a constant (e.g., zero). In this case, equation (10.6) can be rewritten as follows:

$$x = \frac{A_1X + A_2Y + A_3}{A_7X + A_8Y + 1}, \quad y = \frac{A_4X + A_5Y + A_6}{A_7X + A_8Y + 1}. \tag{10.7}$$

This is called a 2-D projective transformation, which is a special case of the direct linear transformation (DLT). There are 8 coefficients in this model and there is a need of at least four control points for the computation of coefficients. This model becomes an approximate model when there are height variations on the earth's surface. Indeed, if very high accuracy is not a requirement, then there is no need to do any image corrections and such an approximate model is very desirable if the terrain is very flat and/or accuracy is not an issue. Similarly, the Z value in equation (10.5) could also be ignored.

In the case of a photograph, one frame of an image is taken at one time. For each frame, there are six orientation elements (three rotation and three translation). The elements are the same for all image points (pixels) on the same frame. Therefore, there is a need to compute the coefficients once for each frame. In the case of scanner images, however, image pixels are acquired scan by scan. A frame of an image may consist of thousands of scan lines (see Figure 10.6). In this case, there are six orientation elements and there is a need of at least three control points for the computation of coefficients for each scan line. This is not feasible. The problem could be simplified if the relationship among the scan lines is known. However, this is normally not the situation. In this case, a single (normally second order) polynomial is used to approximate the geometric model transformation for all scan lines:

$$\begin{aligned} x &= a_0 + a_1 X + a_2 Y + a_3 X^2 + a_4 Y^2 + a_5 XY + \cdots \\ y &= b_0 + b_1 X + b_2 Y + b_3 X^2 + b_4 Y^2 + b_5 XY + \cdots \end{aligned} \qquad (10.8)$$

Indeed, in this case, the errors caused by distortions are re-distributed rather than removed by the polynomial function.

With this geometric relationship established, all image points can then be transformed onto a geo-referencing system. As a result of various distortions, the originally gridded pixels (e.g., 25 μm by 25 μm) on the image, like Figure 10.2a, may not be in a grid form anymore after the transformation. The resultant image on the ground (called ground image in this context) could be something like that depicted in Figure 10.2b.

Theoretically speaking, geo-referencing is completed after these two stages, i.e., removal of scale variation and registration to a geodetic system. However, the story is not over because of some practical problems.

In practice, it is inconvenient to work with such irregular ground pixels as those shown in Figure 10.2b. This requires the rearrangement of the ground pixels into gridded, normally squared, pixels. Even when the ground pixels are in regular grid form already, it is also possible that sometimes one may want to generate an orthoimage with a pixel size different from that of the ground pixels, thus also requiring the rearrangement of the pixels. This process is called resampling in image processing but is called radiometric transformation in this context and is discussed in section 10.2.2. Procedures for orthoimage generation is discussed in more detail in section 10.4.

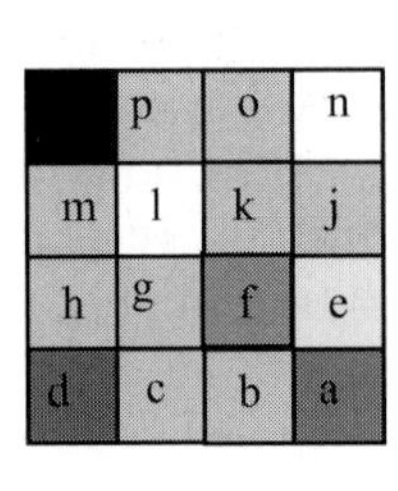

(a) Pixels of a part of an image before geo-referencing

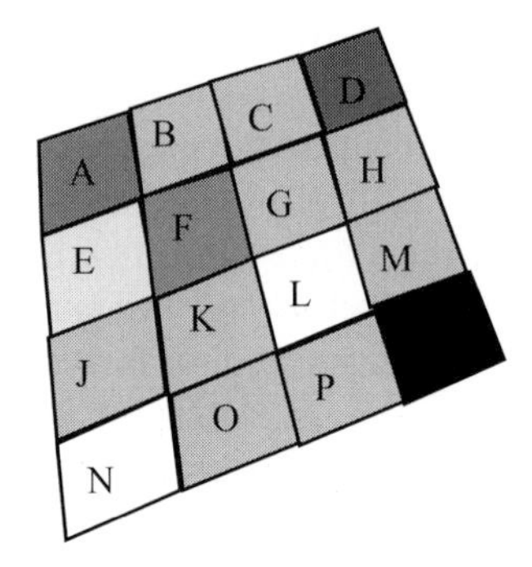

(b) Corresponding ground pixels after geo-referencing

Fig. 10.2 An image in grid form is transformed to a geo-referencing system

10.2.2 Radiometric Transformation

In practice, orthoimages are also arranged in a matrix form, normally with square pixels, each representing a certain ground size (e.g., 5 m × 5 m). The principle of radiometric transformation is illustrated in Figure 10.3. In this figure, with a given pixel size of the resultant orthoimage, a template can be produced to cover the area of interest or the whole area of the ground image. The template pixels are numbered from 11 to 55. The process of radiometric transformation is to obtain a colour (grey value) for each template pixel from the corresponding ground pixels. This is called resampling in other image processing literature.

As shown in Figure 10.3a, there is no one-to-one mapping between the template pixels and ground pixels. One template pixel may be occupied by more than one ground pixel. For example, pixel 22 of the template is occupied by parts of four pixels of the ground image. Alternatively, one ground pixel may cover more than one template pixel. For example, ground pixel B covers parts of template pixels 12, 13, 22, and 23. It is also possible for one template pixel to cover a number ground pixels if the pixel size of the template, which represents the scale of the resulting image, is very large. Then the problem arises of assigning colours to the template pixels.

There are two approaches. The first one is to take the image pixels for primary consideration. An image pixel is transformed onto the ground to form a ground pixel as shown in Figure 10.2. The colour of this ground pixel is then distributed to the template pixels which are covered or partially covered by this ground pixel. For example, image pixel c in Figure 10.2a become ground pixel C in Figure 10.2b after the geometric transformation and the colour of the ground pixels will be distributed to template pixels 12, 13, 22, and 23 in Figure 10.3a, using some interpolation methods. This is regarded as the image-primary approach in this chapter and is also regarded as the forward or direct method by other authors. This is widely used for geo-referencing of images acquired using linear scanners. Figure 10.3b shows the assignment of colour (grey) to each of the template pixels shown

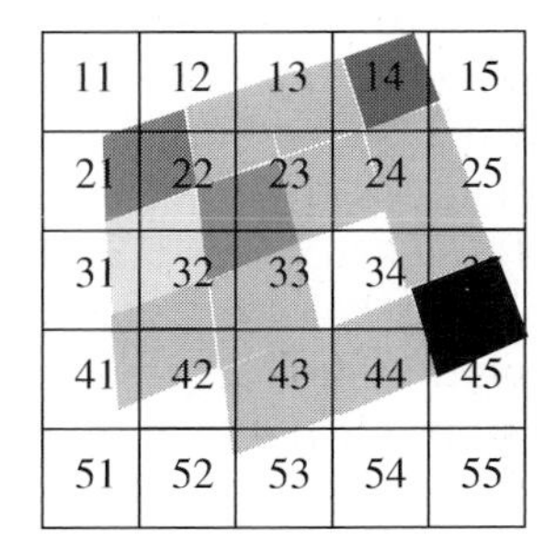

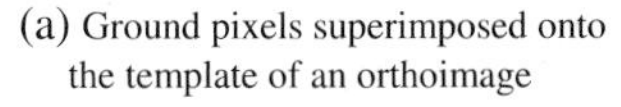
(a) Ground pixels superimposed onto the template of an orthoimage

11	12	13	14	15
21	22	23	24	25
31	32	33	34	35
41	42	43	44	45
51	52	53	54	55

(b) The resultant orthoimage using nearest neighbour interpolation (of grey)

Fig. 10.3 Geo-referencing images: radiometric transformation

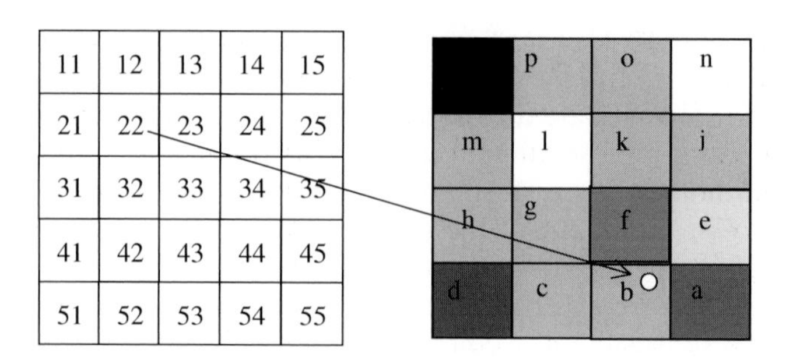

(a) Template of orthoimage (b) The image to be geo-referenced

Fig. 10.4 The space-primary approach for radiometric transformation. The centre of template pixel 22 is projected onto image plane to form a "float" image point (the white dot at the upper/right corner of image pixel b). The colour of this "float image" is obtained from either image pixel b or the neighbouring image pixels a, b, e, and f. The colour of template pixel 22 is then assigned the value of the "float" image point

in Figure 10.3a, to produce an orthoimage. The method used for obtaining the colour (grey) for the template pixels in this case is called "nearest neighbour," i.e., to take a value (of colour) from the nearest (or closest) pixel. A detailed discussion of this interpolation method will follow later in this section.

In the image primary approach, one is concerned with the distribution of the colour of each image pixel to the relevant template pixels of the orthoimage. The line of thinking is from image to space. Alternatively, one can also think in a reverse direction, i.e., from space to image. In the space primary approach, template pixels of an image are taken for primary consideration. The centre of each template pixel is projected (i.e., transformed) onto the image plane to form a "float" image point. The colour of this "float" image point is then obtained from its neighbouring image pixels, using an interpolation method. Figure 10.4 illustrates this approach. In this figure, the centre of template pixel 22 is projected to the image plane to form a "float" image point, i.e., the white dot at the upper/right corner of image pixel b. The colour of this "float image" is obtained from either image pixel b or the neighbouring image pixels, i.e., a, b, e, and f. The colour of this "float" image point is then passed to template pixel 22. This line of thought is also referred to as the backward or indirect methods by other authors. This is widely used in geo-referencing photographs.

Now comes the problem of interpolating colour from neighbouring pixels. Theoretically speaking, any interpolation model can be used for this purpose. According to sampling theory, the ideal function is called the "delta" function, i.e., a sharp rectangular pulse, and the bicubic function is the best approximation. However, due to the larger number of coefficients in this model, it is not widely used. In practice, bi-linear interpolation is the most popular model. Nearest neighbour interpolation is favoured in some cases.

Indeed, these methods are also applicable in the case of interpolation of elevation values from a DTM for a ground point, which is required in the geometric transformation of the geo-referencing process, as specified by equations (10.4) to (10.7). Therefore, in the discussion of this section, a more general term – height – is used to refer to either the elevation or the colour.

"Nearest neighbour" is the simplest method. It takes the required attribute (e.g. height) from the pixel's nearest neighbour. The principle is illustrated in Figure 10.5a. In this figure, the heights for Points A and B need to be interpolated from their neighbours – Points 1 to 8. Using the "nearest neighbour" methods, the height of Point 4 will be assigned to Point A as Point 4 is closest to Point A. Similarly, the height of Point 5 will be assigned to Point B. The mathematical function is as follows:

$$h_A = h_i,\ if\ d(A,i) = \min[d(A,1), d(A,2), \ldots, d(A,i), \ldots, d(A,N)] \quad (10.10)$$

where h represents an attribute of the points, i.e., height or colour; A is the point whose attribute is to be interpolated; $d(A,i)$ is the distance between Point A and Point i; and N is the total number of reference points, i.e. eight in Figure 10.5a.

Bilinear interpolation is another popular method. It is used for interpolation from gridded data points. Figure 10.5b shows the principle of the method. In this figure, Points (i, j), (i+1, j), (i + 1, j + 1), and (i, j + 1) are the reference points located at grid nodes; A is the point whose attribute (e.g., height) needs to be interpolated. In order to obtain the attribute (height) for A, the attributes (heights) of two intermediate points – 1 and 2 – are interpolated linearly from points (i, j) and (i, j + 1) and from points (i + 1, j) and (i + 1, j + 1), respectively. Then the attribute (height) of Point A is interpolated from Points 1 and 2. Of course, one can interpolate two intermediate points in the other direction first, however, the result will be identical.

The mathematical formula is as follows:

$$\begin{aligned} h_A = {} & h_{i,j} + \Delta x(h_{i+1,j} - h_{i,j}) + \Delta y(h_{i+1,j} - h_{i,j}) \\ & + \Delta x \bullet \Delta y(h_{i,j} - h_{i+1,j} - h_{i,j+1} + h_{i+1,j+1}). \end{aligned} \quad (10.11)$$

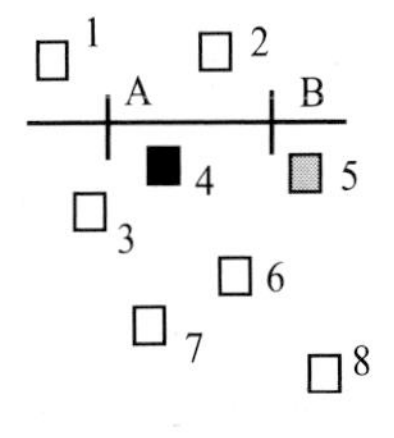

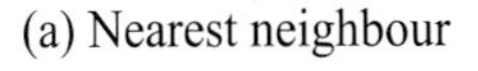
(a) Nearest neighbour

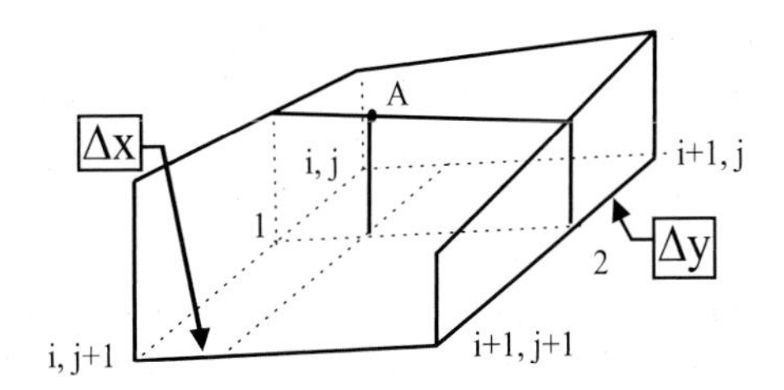

(b) Bilinear interpolation

Fig. 10.5 Interpolation: nearest neighbour and bilinear methods

(a) A portion of a scanned imagery before rectification

(b) The same image after rectification and superimposed with digitized lines

Fig. 10.6 An example of an orthoimage from scanned imagery

An example of a geo-referenced image using the bilinear interpolation model is shown in Figure 10.6.

10.2.3 Joining Adjacent Images – Image Mosaicing

After geometric and radiometric transformations, an orthoimage from a single frame (or a strip) can be produced. In many cases, however, it is necessary to combine adjacent frames (or strips) to form a larger orthoimage. This process is called "mosaicing".

One could imagine that adjacent images could match well if there were no errors involved in the imaging and then the geo-referencing process. In this case, the mosaicing process would be to simply cut down those overlapping areas. However, errors are inevitable and the overlapping area of two images would not match exactly if two images were geo-referenced separately. There are two solutions to the problem. The first solution is to adjust one image relative to the other. The second one is to consider the block of images together. In both cases, control points in overlapping areas need to be employed. This is the geometric part of mosaicing and is called geometric mosaicing.

Fig. 10.7 A mosaicked orthoimage from a seamless combination of two trips and superimposed with digitized linear features

Sometimes, two neighbouring images may have been taken at different times. One may appear much brighter than the other. This phenomenon can be identified by examining features in overlapping areas. In this case, the brightness of the two images should also be normalised. This is the radiometric part of mosaicing and called radiometric mosiacing in this context. An example of a mosaiced image is shown in Figure 10.7.

After these three section, the principle of geo-referencing images has been introduced. A more detailed discussion of the procedure, however, will take place in section 10.4.

10.3 Measurement from Single Image

From a single image, some geometric information and thematic information can be acquired. Geometric information is usually acquired through measurement while thematic information is acquired by interpretation. Both measurement and interpretation can be carried out either manually or automatically. In the automated mode, the measurement is carried out by edge detection algorithms, while interpretation is by classification algorithms. In this chapter, however, only measurement will be discussed and interpretation will be explained in Chapter 12.

10.3.1 Measurement Mode and Image Format

Images could be either in digital form or in analogue form. In a digital image, measurement can be done on screen and the unit of measurement is a pixel. The results of measurement are sets of coordinates from which geometric information, such as distance, dimension, and height, could then be computed. If the image is an orthoimage, then a pixel represents a physical dimension on the ground, e.g., $3\,\text{m} \times 3\,\text{m}$. The results of measurement could be coordinates in an already geo-referenced system. If the image has not been geo-referenced yet, then there are distortions in the image position and all image coordinates need to be transformed to a geo-referencing system. The principle of such a transformation is exactly the same as that for orthoimage generation. If an image is in analogue form, one could follow either of three ways, namely:

a) Way One: digitise those features in a way similar to the digitisation of analogue maps and then transform all the digitised points to a geo-referencing system.
b) Way Two: do direct measurement on the analogue image to obtained distance and then to compute other geometric information (e.g., height difference) from this distance measurement.
c) Way Three: first convert the image into digital form using a scanner and then do measurement on the digital images.

10.3.2 Measurement of Coordinates on Single Image

Measurement of coordnates on digital images means to digitise features from a digital image. The process of such measurement is so simple that no further discussion is needed. If the image is an orthoimage already, the measured coordinates are in a geo-referencing system. However, if the image is not an orthoinage, then the measured data need to be geo-referenced by applying a geometric transformation, as discussed in section 10.2.1. In this case, the image-primary approaches should be used. The actual process is to:

a) measure control points;
b) compute parameters for the geometric transformation model (section 10.2.1) employed; and
c) transform the coordinates of every measured image point to the geo-referencing system using the transformation parameters computed at Step (b).

In this case, if high accuracy is not of importance and/or the terrain is flat, then the Z values could be set to a representative value (e.g., the average of the control points).

The digitisation of an analogue image is similar to that of analogue maps and the principles and procedure discussed in Chapter 5 apply here. After measurement, all points should be transformed to a geo-referencing system as for the measured data from digital images.

Indeed, the procedure is quite similar to the digitisation of an analogue image and the only difference is that different mathematical models are used to perform the geometric transformation. In the case of images, the transformation is between 3-D and 2-D surfaces and the models described in section 10.2.1 should be used. On the other hand, in the case of maps, the transformation is between 2-D and 2-D surfaces and an affine transformation is the most popular.

10.3.3 Manual Measurements on Analogue Image

Nowadays, most of the satellite images have been recorded in digital form already. Attempts have been made to acquire aerial digital images using digital cameras, but they have not been very successful [Thom, 1993]. Therefore, aerial photographs are the only analogue images that are still widely used.

Aerial photographs for measurement are normally taken by a professional camera. As discussed in Chapter 9, some information, such as the scale, calibrated focal length, flying height, and so on, is recorded. On an aerial photograph, there are some marks, called fiducial marks, at the four corners and/or at the four sides. These marks can be used to determine the centre of the photograph.

If the distance, e.g., oa in Figure 10.8, is measured, then its ground distance can be computed by multiplying a scale factor, i.e.,

$$D = S \times d = \frac{H}{f} \times d \tag{10.12}$$

where d is the distance measured on the aerial photograph, H is the flying height, f is the focal length of the camera, 1:S is the scale of the photograph, and D is the ground distance. However, this is only an approximate value since there are various distortions in the photograph.

The height of features can also be estimated using measurement from an analogue photograph. For example, at the upper left corner of Figure 10.8, the bottom and top of a building are marked as a′ and a, the height of this building could then be estimated by measuring the distances – aa′ and oa. However, all measurement must be done along a radial from the centre of the photograph because the distortions are radiating from the centre. The geometric relation between the actual building and the image building is illustrated by Figure 10.9. The height of the building, i.e., h in Figure 10.9, can be expressed by the following formula:

$$h_a = \frac{\Delta r}{r_a} \times H = h_a = \frac{\Delta r}{r_a} \times S \times f \tag{10.13}$$

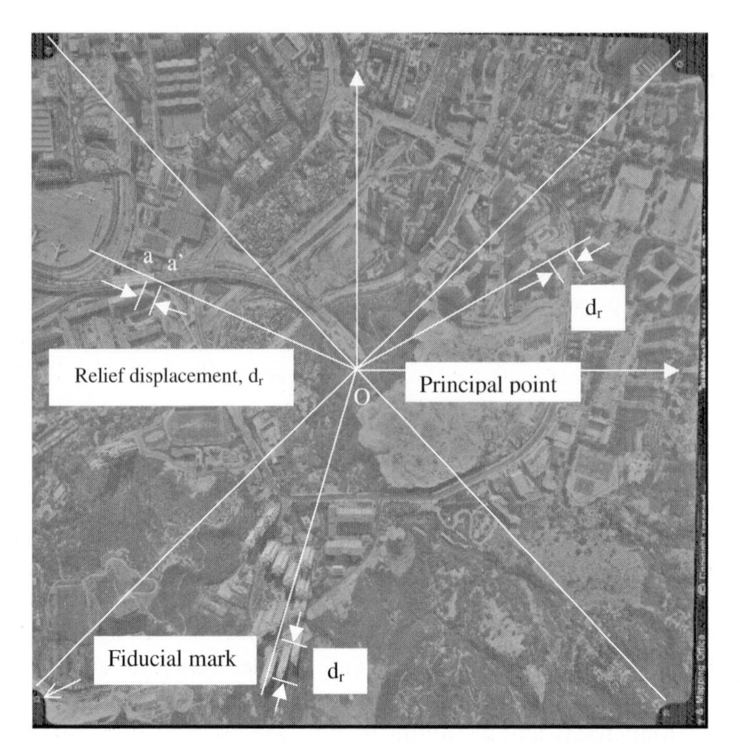

Fig. 10.8 Measurement on an analogue photography

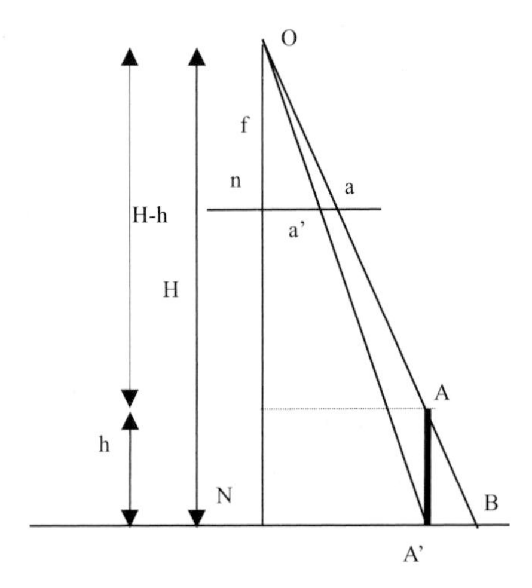

Fig. 10.9 Height difference and image displacement

where, r_a means the radial distance of image point a from the centre of the image, Δr is the radial distance between the top and bottom of the building on the photograph, H, f, and S have the same meanings as in equation

(10.11); and h_a is the estimated height of image point a (from a′). Again, this is only an approximate estimate.

10.3.4 Automated Measurement from Digital Image by Edge Detection

Automated measurement is achieved by a technique called edge detection. This type of technique is based on the dissimilarity of images. The principle is illustrated in Figure 10.10. Figure 10.10a shows an image with a clear edge, and Figure 10.10b shows the grey-level profile along a horizontal scan line. Figure 10.10c shown the edges detected after applying an edge-detection algorithm.

An edge detection is normally based on the idea that there will be a big change in the gradient (or slope) of the grey-level plot (see Figure 10.10b). The gradient at a pixel is computed using pixels within a small window around it such as a 3×3 mask. For convenience of expression, the coefficients of the formulae for the computation of gradients are set in a matrix form, resulting in a special name – edge-detection operators. A number of such operators have been designed and can be found in textbook dealing with image processing. The most popular one is the Sobel operator as shown in Figure 10.11.

In the actual edge-detection process, a mask will be moved over the image (Figure 10.12a) pixel by pixel to produce a gradient value for all pixels. As one can imagine, there will be variations of gradient values among the pixels but only those pixels with large gradient values are a real edge (Figure 10.12b). Then a threshold is set to retain only those pixels that have gradient values larger than the threshold (Figure 10.12c). This works well when an edge is as sharp as that shown in Figure 10.10a. However, as shown

(a) A sharp edge

(b) A grey-level profile along a horizontal scan line

(c) Results of edge detected

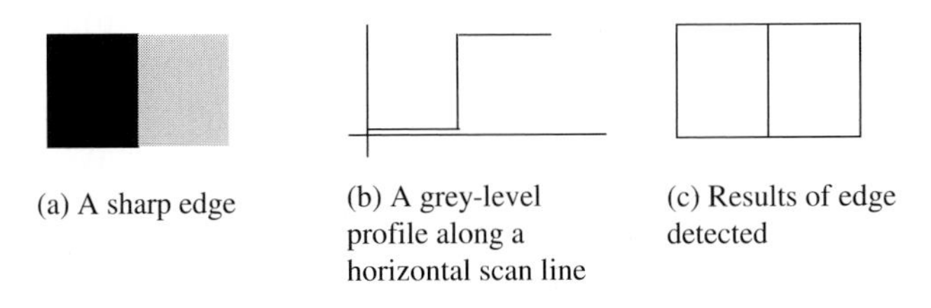

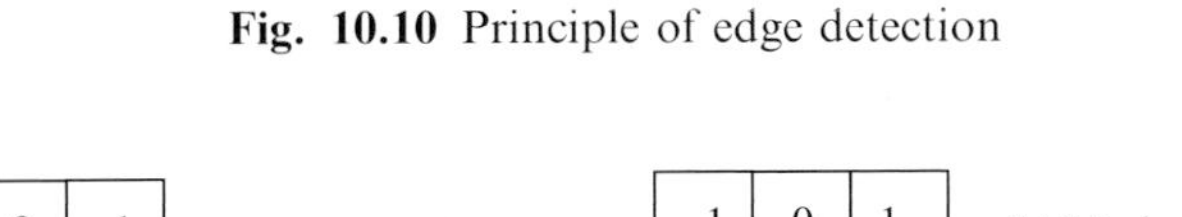

Fig. 10.10 Principle of edge detection

-1	-2	-1
0	0	0
1	2	1

(a) Mask used to compute the gradient in horizontal (x) direction

-1	0	1
-2	0	2
-1	0	1

(b) Mask used to compute the gradient in vertical (y) direction

Fig. 10.11 Sobel operators for edge detection

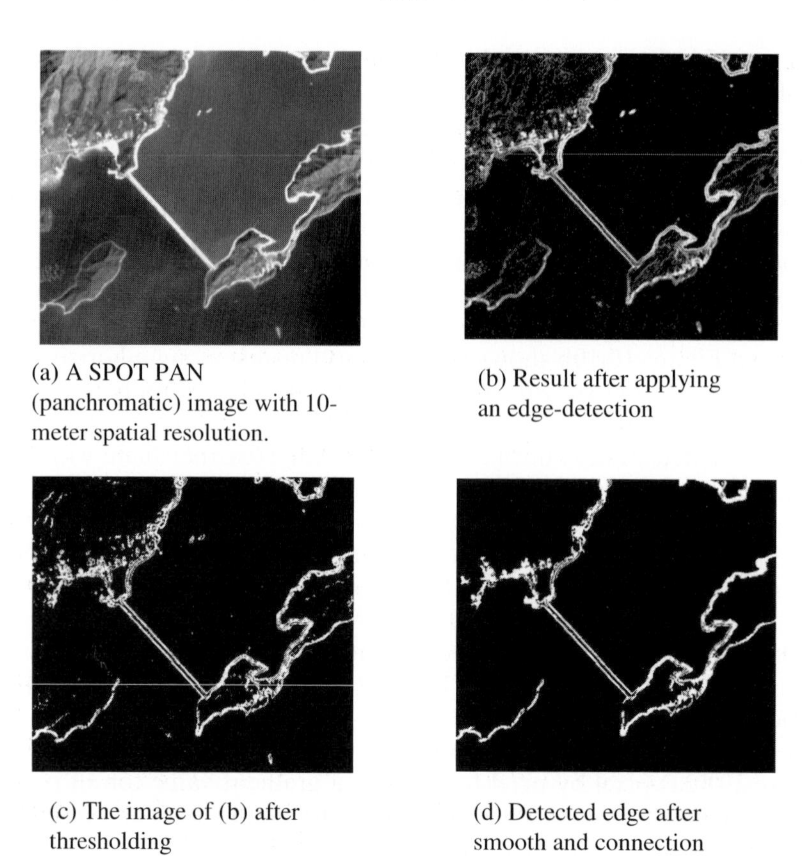

(a) A SPOT PAN (panchromatic) image with 10-meter spatial resolution.

(b) Result after applying an edge-detection

(c) The image of (b) after thresholding

(d) Detected edge after smooth and connection

Fig. 10.12 The steps of edge detection

in Figure 10.12c, the result of edge detection is not as perfect as one expects because of many reasons. Therefore, some post-processing is required such as a smoothing filter to remove small "pseudo-edges" and a connecting algorithm to connect edges separated by pixels removed in the threshold operation (Figure 10.12d). However, a discussion of post-processing lies outside the scope of this chapter and those readers who have an interest in this topic could find information in textbooks such as that by Gonzalez and Woods [1992].

10.4 Procedure for Geo-Referencing Image and Measured Data

As discussed in section 10.2.2, there are two approaches to performing geo-referencing. The first one is called the image-primary approach because a line of thought from image to ground is followed. The other is called the space-primary approach because a line of thought from ground to image is followed. The procedures for both approaches will be discussed in this section 10.4.1.

10.4.1 A Comparison of Image-Primary and Space-Primary Approaches

The space-primary approach starts from the known ground coordinates (X,Y). With given (X,Y), the Z value can be interpolated from a DTM using bilinear interpolation and/or other methods (see section 10.2.2). With known ground position (X,Y,Z), the ground point can be projected to the image plane using the geometric transformation models discussed in section 10.2.1, resulting in a position (x, y) on the image plane. If the image is an aerial photograph, such a corresponding position on an image is always there as the parameters of the geometric transformation models are identical for all image points. However, in the case of scanned images, there should be a set of parameters for each individual scan line. From a point on the ground, one really does not know to which scan line this point will correspond. Therefore, the space-primary approach is not a good approach for scanned images if the rigorous geometric transformation is to be employed. However, if the differences among scan lines are to be neglected so that a polynomial function is used as an approximate model for the geometric transformation, then the space-primary approach is still very useful.

On the other hand, in the image-primary approach, one starts from an image point (x, y) and wants to determine the corresponding ground position (X,Y). This can be done only when the Z value for the (X,Y) are known. This is the difficult part of the story. To know the Z value, one must know the (X,Y) position first and then apply an interpolation process to a DTM. But, as indicated in equation (10.4), one must know the Z value in order to obtain an accurate (X,Y). To resolve this dilemma, two methods can be employed. The first is to form a terrain surface using an existing DTM and then obtain the (X,Y,Z) position of a ground point simultaneously by intersecting the light ray with the DTM surface. The other method is to employ an iterative solution. The principle of the iterative process is illustrated in Figure 10.13. At first, using the datum (reference), the corresponding ground X (i.e., X_1 in Figure 10.13) and Y coordinates of an image point (pixel) are computed using a geometric transformation model (section 10.2.1). A new height value for this can be obtained for this point through

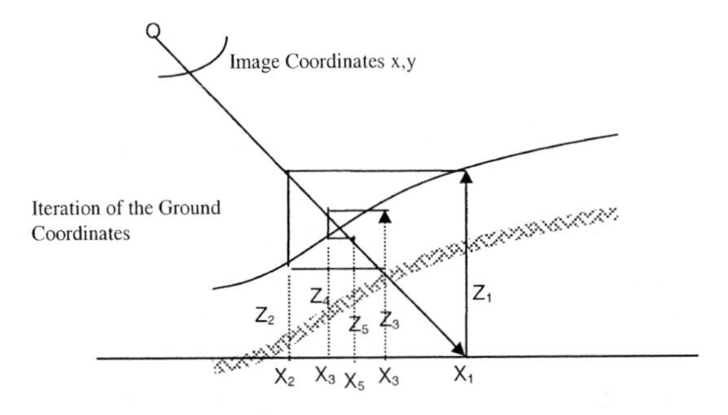

Fig. 10.13 Iterative determination of height and position of a ground point

interpolation from a DTM (Z_1 in Figure 10.13). Then this new height is used as Z in the geometric transformation model to compute new X (X_2 in Figure 10.13) and Y coordinates for the point (pixel). After a few iterations, accurate results can be obtained (Z_5 in Figure 10.13). Experience gained in this study reveals that after two iterations, good results can be achieved easily. This image-primary approach is popular for the rigorous geo-referencing of scanned images.

10.4.2 Object-Primary Approach for Orthoimage Generation

In section 10.4.1, the line of thought for the space-primary approach has been briefly outlined. Then, the following problems arise:

a) how to set a ground template;
b) how to project every template pixel to the image plane; and
c) how to interpolate a colour for each template pixel.

To set a template for the resulting orthoimage, the possible ground coverage of the image and the desirable ground pixel size need to be known. To project every template pixel to the image plane, the coefficients of the geometric transformation model need to be obtained, such coefficients expressing the relationship between the image points and the corresponding object points on the ground. In these cases, a set of ground control points is required. The procedure is:

Step 1: Identify and measure a few ground control points. The actual number required is at least half the number of coefficients in the mathematical models because two equations for the model can be set for each ground control points.
Step 2: Compute the coefficients in the geometric transformation models using the method of least square, which was briefly discussed in Chapter 3. A more detailed discussion of the method of least square can be obtained from text books (e.g., Mikhail [1976]).
Step 3: Transform four corners of an image onto the ground so as to obtain ground coverage of the orthoimage, i.e., the maximum and minimum X and Y coordinates.
Step 4: Determine the pixel size of the output orthoimage. The actual size should be determined according to the scale of the output image. In any case, the pixel size should not be smaller than the average ground pixel size.
Step 5: Set a template for the output image, i.e., compute the number of rows and columns, the coordinates at the centres of all template pixels.
Step 6: Estimate the height value for each pixel in the template by interpolation from a DTM.

Step 7: Project each template pixel onto the image plane using the coefficients obtained in Step 2 to locate the (x,y) position of the point in the image. The x and y coordinates are in real (or float) format (but not in integer).

Step 8: Apply a radiometric transformation to the (x,y) so as to get a colour from its neighbouring image pixels.

After Step 8, each pixel in the template is assigned a colour and an orthoimage is then obtained. For a very flat area, Step 6 can be omitted. This method is widely used when the source images are photographs.

10.4.3 Image-Primary Approach for Orthoimage Generation

In the image primary-approach, the steps after Step 5 are as follows:

Steps 1–5: The same as for space-primary approach.

Step 6: Project image pixels one by one to the ground using an iterative process, as shown in Figure 10.13, to obtain ground pixels.

Step 7: Obtain a colour for the template pixel by interpolation using its neighbouring ground pixels. Some alternative methods for this type of radiometric transformation have been discussed in the literature [Ebner et al., 1980; Zhang et al., 1994].

After Step 7, each pixel in the template is assigned a colour and an orthoimage is then obtained. This method is used when the source images are scanned images.

10.5 Summary

In this chapter, it has been revealed that there are various distortions with images and that a geo-referencing process needs to be applied in order to make images as useful as maps or to make data measured from images useful. In summary, two main topics were covered in this chapter: geo-referencing images (orthoimage generation) and taking measurement information from single images and geo-referencing that measured data.

In geo-referencing images, two key steps need to be taken. The first step is to establish the relationship between the coordinates (x,y) of image points in the image coordinate system and the coordinates (X,Y,Z) of the corresponding object point on the ground in a geo-referencing system. Then every image point is transformed into ground coordinates in the geo-referencing system, with the aid of a few known ground points (called control points). This step is called geometric transformation. The next step is to transform the colour (or grey value) of these points to corresponding positions of the ground template of an orthoimage. This step is a process called radiometric transformation.

In the geometric transformation, image coordinates need to be corrected for distortions if high accuracy is required for the resultant orthoimage. Among these distortions are lens distortion, earth's curvature, atmospheric refraction, relief, and tilt. In the mathematical models for the geometric transformation, however, the last two – tilt and relief that are the most significant ones – have been taken into consideration. Therefore, in GIS practice, such image correction is not a big issue.

Similarly, if there is no need for high accuracy, approximate mathematical models can be used for the geometric transformation. In practice, polynomials are widely used as the approximate models for geo-referencing satellite images obtained by scanners.

In radiometric transformation, the most widely used models are nearest neighbour and bilinear interpolation. The former can retain the sharpness of the images but may create position errors as large as half a pixel. The bilinear method will guarantee better position accuracy but will cause a slight degradation of image sharpness.

There are two approaches for geo-referencing images, i.e., image primary and space primary. The former is more convenient i when scanned images are to be geo-referenced using rigorous geometric transformation models. The later is more convenient when photographs are to be geo-referenced and when an approximate geometric transformation model is to be used.

After an image is geo-referenced, the resultant image can be used (a) as backdrop to other spatial information in GIS; (b) as a source from which measurement of spatial objects and extraction of other spatial information can accomplished; and (c) to produce image maps with superimposed contour lines. For these reasons, orthoimages have been regarded as the most important type of spatial data in a national geospatial data infrastructure.

Geometric information can be measured from single images. Such information may be distance, coordinates, and height differences. Measurement could be accomplished either manually or automatically. If the images used for measurement are orthoimages, then the resultant information should be in a geo-referencing system already. However, if an ordinary image is used, then each measured point needs to be geo-referenced. The procedure is similar to the image-primary approach for orthoimage generation.

11 Geometric Data from Images

Bruce King

11.1 Introduction

Chapter 10 showed how to geo-reference single images and how the position of features may be determined from those images. There are several limitations to these techniques if "high" precision and accuracy is required for the position of features. Implicit in single image methodologies for obtaining height is the need to have an elevation model. But we know that height information already exists in an aerial photograph in the form of relief displacement. This chapter introduces the concepts of photogrammetry – the extraction of geometric information from images – and shows how it is used to extract more reliable position data from overlapping images and blocks of images. Thus it will lead to the geo-referencing of stereo (and multiple) images.

Early efforts to extract three-dimensional information from photographs produced methodologies that separated planimetric position and height. These methodologies are still practiced today in elementary mapping techniques, and are useful as an aid to understanding more advanced geo-referencing methods. More advanced methodologies enable the extraction of three-dimensional information from a single, simultaneous observation. Both approaches rely on the concept of stereoscopic vision. In addition, the processes have been transformed from mechanical and analytical operations into digital operations that have automated the data extraction task.

Both the separate and simultaneous methods will be discussed and their limitations and applications outlined. Before this, it is necessary to have an appreciation of stereoscopic vision and how it is applied in photogrammetry.

11.2 Stereoscopic Vision

When we look at an object at a particular distance, our eyes are simultaneously focused on and convergent upon the object. The angle of convergence is called the *parallactic angle*. This is easily demonstrated by holding a finger in front of your nose at arm's length. As you move your finger towards your nose and follow it with your eyes, they converge to greater angles. This is a change in parallactic angle. Objects at different distances from us are perceived through different parallactic angles. These concepts are illustrated in Figure 11.1. Thus, one way to measure distance is to measure the parallactic angle.

Indeed this is fundamentally how we estimate distances. The brain converts parallactic angle into an assessment of distance. The eye is very sensitive to changes in parallactic angle and so can discern changes in depth quite accurately.

Because of the distance between our eyes (our eye base), we see an object from two different points of view. Consequently, due to the parallactic angle, the image of an object point will fall on different relative locations of the

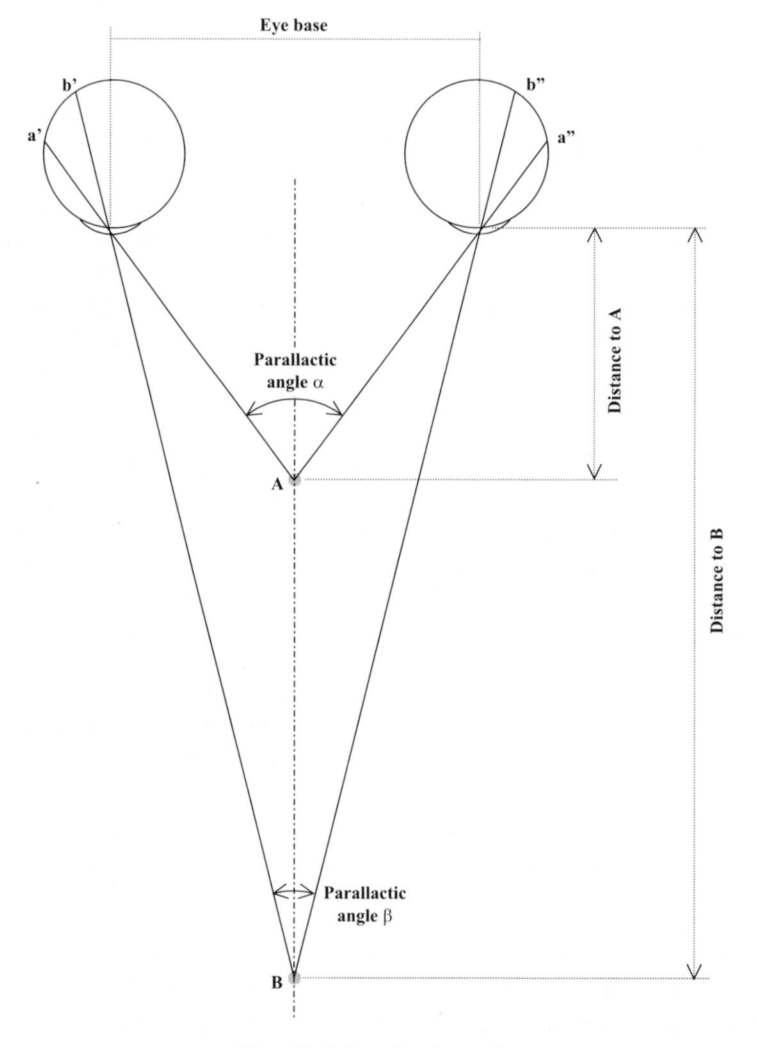

Fig. 11.1 Parallactic angle

retinas of our left and right eyes. Points at different distances from our eyes will have a different difference in relative location. This is illustrated in Figure 11.1 by the different locations of points **a**′, **a**″, **b**′ and **b**″. The change in position of the image points is called *parallax*. Parallax can also be demonstrated by holding a finger in front of your eyes and seeing how its position changes with respect to objects in the background when viewed alternately with each eye. So changes in parallactic angle result in parallax.

Instead of measuring parallactic angle, distance (depth) can be determined by measuring parallax. This method of measuring depth leads to an accurate method of measuring height in overlapping photographs.

11.3 Stereo Photographs

Chapter 9 presented two important concepts that are applied here. First, a photograph is a perspective projection and exhibits relief displacement. Second, in aerial photography blocks of overlapping photographs are obtained as the aeroplane travels across the sky over the terrain. Relief displacement is proportional to the height of a point (above the mapping datum) and its distance from the ground nadir point. When examining conjugate points (the same image point on two images) in overlapping photographs, it will be seen that the relief displacement changes. The successive overlapping photographs may be considered to be acting the same as if our eyes were at the corresponding locations of their perspective centres.

Relief displacement occurs in a radial direction from the ground nadir point (also the principal point for vertical photographs), but it can be divided into two components. One component is parallel to the camera base and one is perpendicular to the camera base. The difference in the relief displacement parallel to the camera base is the same as the difference in the image positions of points **A** and **B** as seen in Figure 11.1 and this is called *x-parallax* (or *stereoscopic parallax*). The difference in position perpendicular to the camera base is called *y-parallax*. Y-parallax must be minimised before clear stereoscopic viewing of a stereopair can be achieved.

If, instead of looking at an object directly, we look at two overlapping photographs taken from different locations (such as those obtained from an aircraft), we can use the measurements of x-parallax to determine the distance from the cameras to the objects that appear in the photographs. Such a pair of photographs is called a *stereopair* and a sample is shown in Figure 11.2.

It is important to remember that x-parallax is measured with respect to the camera base – the line showing the path of the camera between successive images. This means that for a strip of overlapping photographs each individual photo may have two camera bases, one relating to the photos on either side.

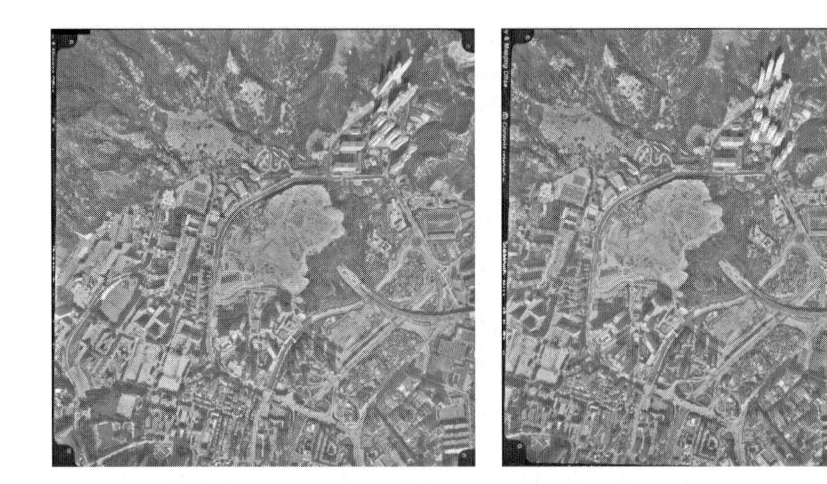

Fig. 11.2 A stereopair of photographs (Survey and Mapping Office, Lands Department, Hong Kong SAR Government)

11.3.1 Viewing a Stereopair

If we now looked at the two photographs, with one eye confined to looking at one photograph and the other eye to looking at the second photograph we would see a three-dimensional model of the object. The perceived object is called a *stereomodel*. Viewing a stereopair is actually quite difficult without the aid of some mechanical device. The simplest device for this task is called a *stereoscope*. A stereoscope set up for viewing a stereopair is shown in Figure 11.3.

There are several other methods for viewing stereopairs; each involves limiting the view of one photograph to one eye. Methods include anaglyph filters, polarised filters, and image alternation. Each of these methods has been implemented in both older analogue systems and current digital systems. Regardless of the viewing method, in order to successfully view a stereopair, the following conditions must be established:

(i) The photographs must be placed in their correct relative positions.
(ii) The photographs must be aligned along the camera base.
(iii) The photographs must be at a spacing suitable for comfortable viewing.
(iv) The viewing or eye base must be parallel to the camera base.

Placing the photographs in their correct relative positions involves identifying the correct order of the photography. This is easily achieved by overlaying the common areas of the stereopair and then separating them in the direction of overlap. Failure to do this step will result in a *pseudoscopic* model – one where the hills look like valleys and vice versa.

As shown in Figure 11.4 for vertical aerial photographs, aligning the photographs along the camera base involves identifying the principal points

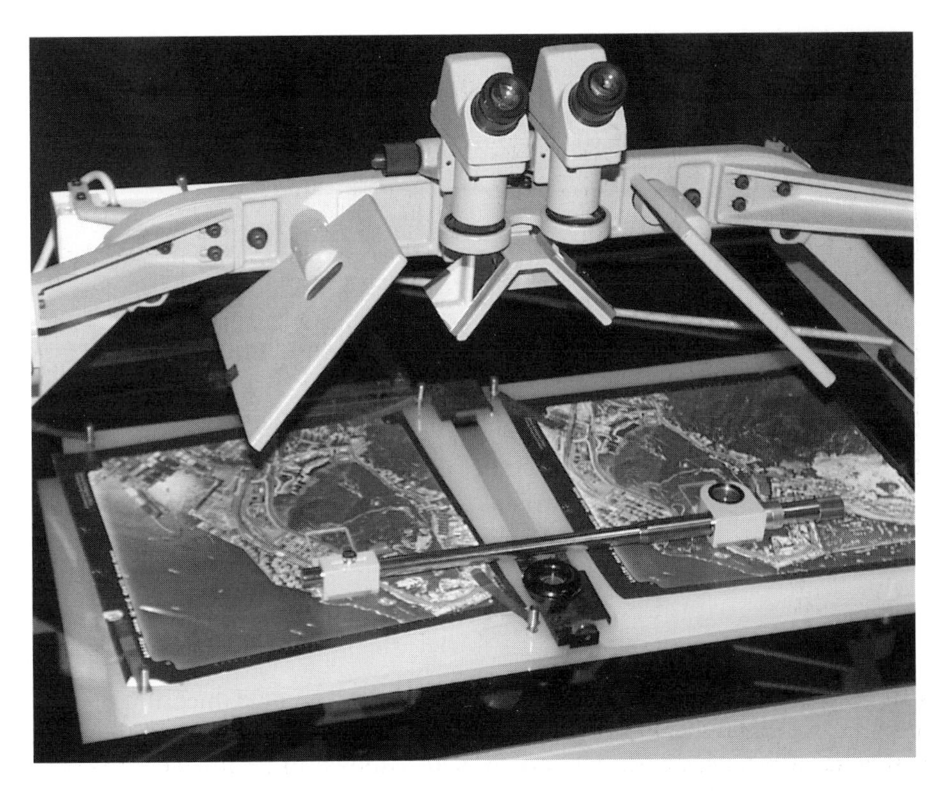

Fig. 11.3 A stereopair, stereoscope and parallax bar

(the ground nadir points) and their conjugate images on the pair of photographs and placing all four points on a straight line. Failure to do this properly is one source of y-parallax, which makes the formation of a stereomodel difficult. Other sources of y-parallax are variation in scale of the photographs either because of a rapid change in flying height or terrain height, and relative tilt of the photographs. Each of these is illustrated in Figure 11.5.

Viewing the stereomodel with the eye base parallel to the camera base is important for correct and accurate stereoscopic vision. This ensures that the viewer correctly interprets the change in relief displacement parallel to the camera base as x-parallax. It does not matter that the left and right eyes of different people viewing the stereomodel are looking at the opposite photographs as would happen if they were looking at a stereomodel from opposite sides of the table.

11.4 Height from Stereoscopic Parallax

Due to the motion of the aeroplane, the difference in position of conjugate points is parallel to the eye base and proportional to the heights of the point

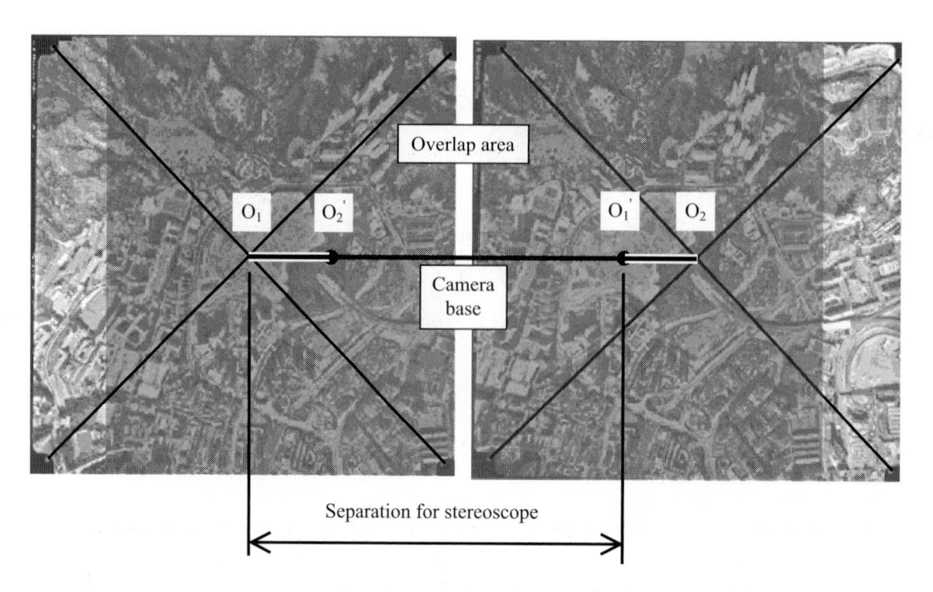

Fig. 11.4 Setting up a stereomodel

and the camera above the mapping datum. If we could measure the x-parallax, we could measure the height of the object. An instrument that can do this is called a *parallax bar*, which is also shown in Figure 11.3.

As shown in Figure 11.6 **A** is closer to the cameras than **C** so image points $\mathbf{a}_1$ and $\mathbf{a}_2$ lie in different relative positions to $\mathbf{c}_1$ and $\mathbf{c}_2$. This is seen more clearly in Figure 11.7 where the two photographs are superimposed on the one perspective centre. The x-parallax is defined in equation (11.1):

$$\begin{aligned} p_a &= x_{a_1} - x_{a_2} \\ p_c &= x_{c_1} - x_{c_2}. \end{aligned} \tag{11.1}$$

The parallax for **A** is larger than that for **C** as **A** is closer to the cameras. There are two basic characteristics to stereoscopic parallax:

(i) parallax of any point is inversely proportional to its distance from the camera;
(ii) parallax due to height occurs only in the direction of flight (x axis).

Height can be computed from parallax measurements. This can be done either directly from parallax observations:

$$h_A = H - \frac{Bf}{x_{a_1} - x_{a_2}} = H - \frac{Bf}{p_a}, \tag{11.2}$$

or in terms of a difference in parallax:

$$h_A = h_C + \frac{(p_a - p_c)(H - h_C)}{p_a}. \tag{11.3}$$

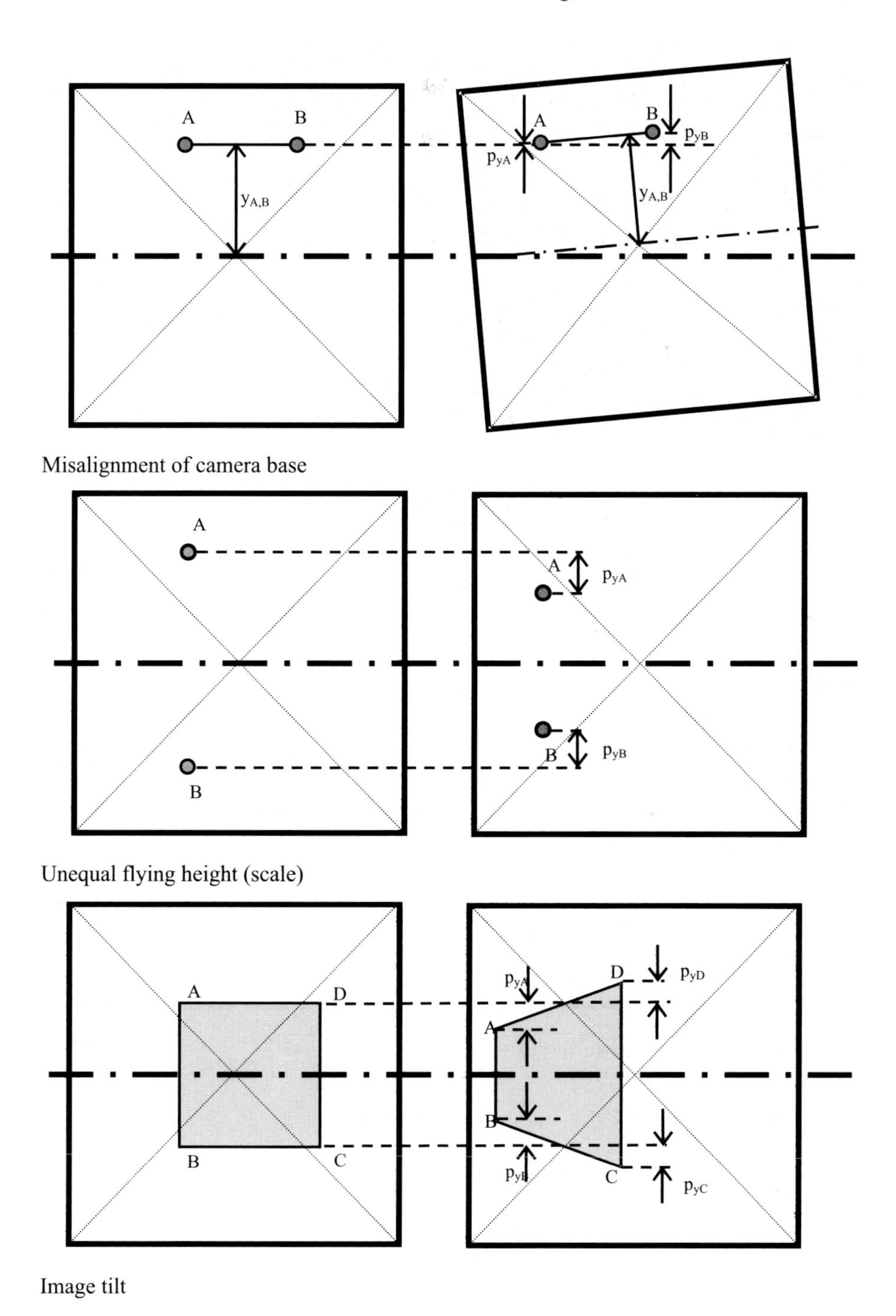

Fig. 11.5 Causes of y-parallax [After Wolf PR (1983)]

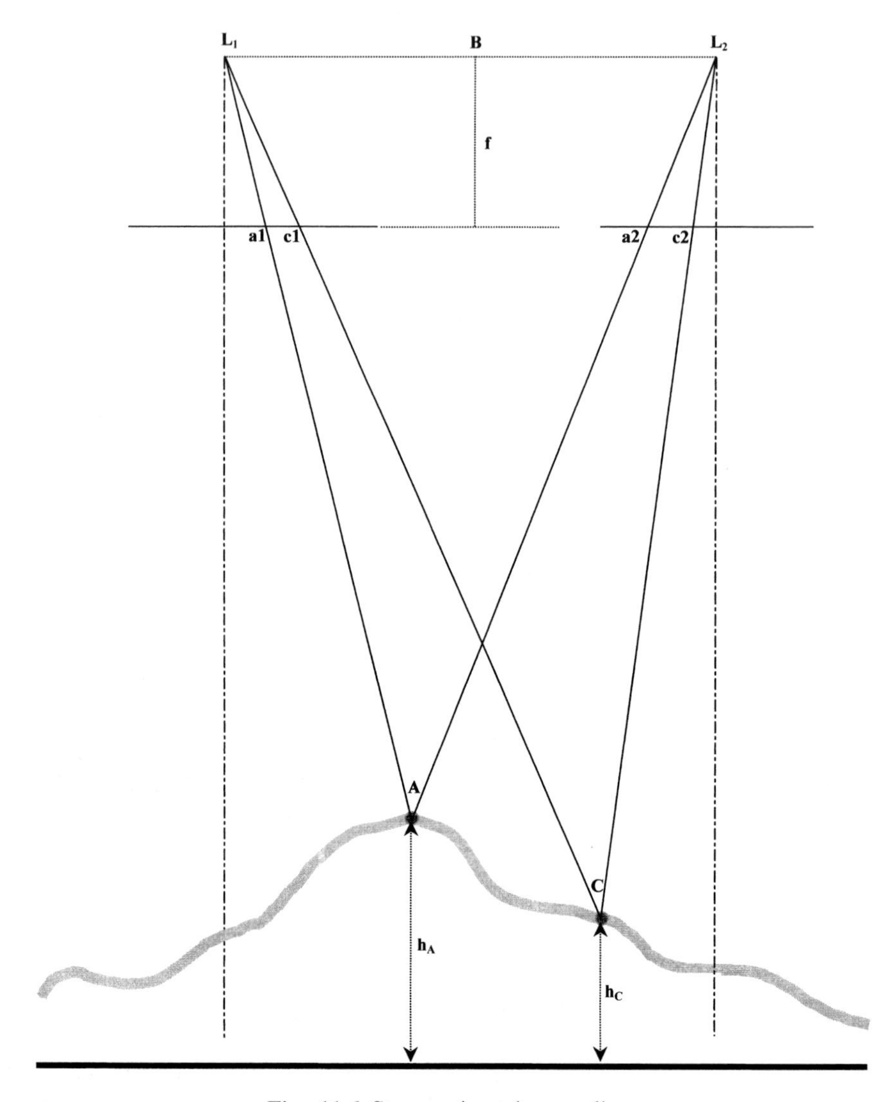

Fig. 11.6 Stereopair and x-parallax

The advantage of using parallax difference is that neither the focal length of the camera lens (f) nor the camera base (B) is required. However, the heights of some points that can be seen in the stereopair are required. These points are referred to as *control points* and contribute significantly to a height determination of greater accuracy than the direct computation from equation (11.2). In equation (11.3) point C is a control point and the height of A is being computed with respect to it. To further reduce the accumulation of errors, several control points may be used over the stereomodel.

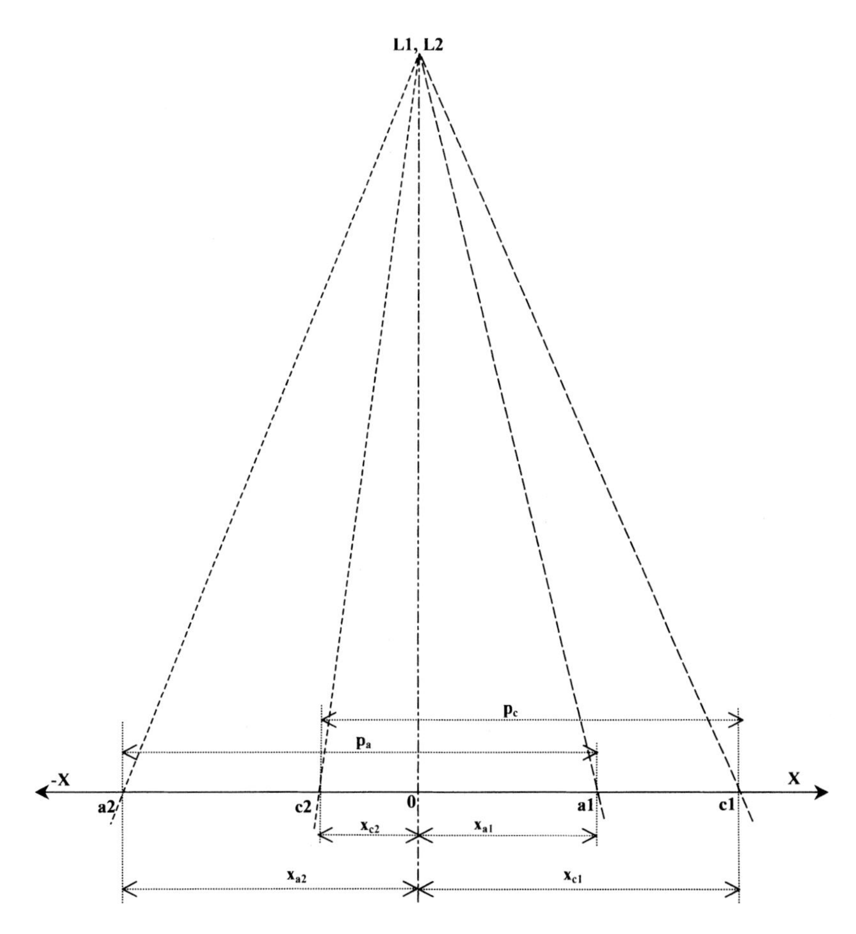

Fig. 11.7 X-parallax

11.4.1 Errors of Parallax Heighting

An evaluation of the two methods above shows that heights computed from parallax differences are more reliable than those computed directly from parallax observations. The sources of errors are both operational (locating the principal points and their conjugates, aligning them to the camera base, and aligning the eye base with the camera base) and statistical (the accuracy and precision of the variables of the functional model used).

11.4.2 Using the Parallax Bar

Before the parallax bar can be used to measure parallax, it is necessary to determine that it is constant. This is because the left end of the bar is adjustable so that a range of parallax can be measured. The relationships between a stereopair and a parallax bar are shown in Figure 11.8.

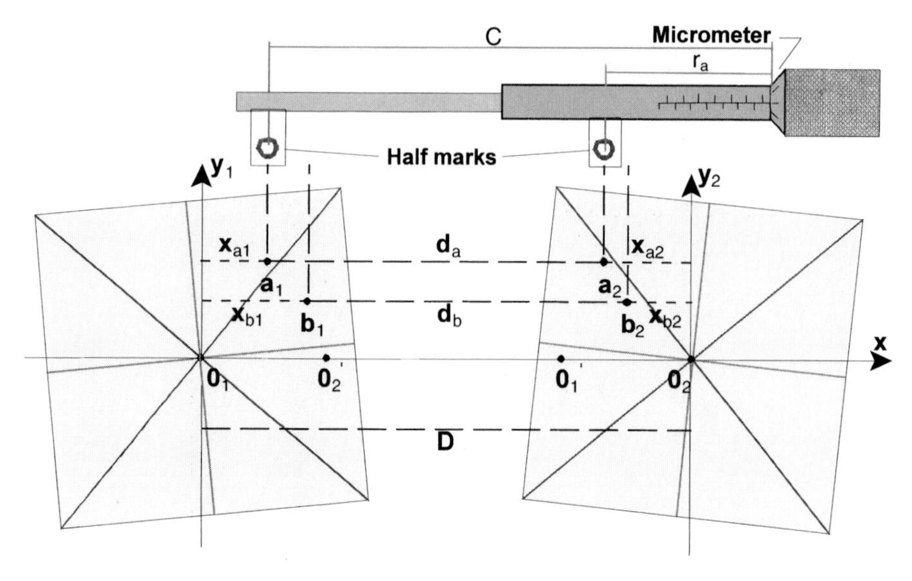

Fig. 11.8 Parallax bar and stereopair

Two half marks are attached by a micrometer, which allows accurate measurement in their change in separation (or parallax). The sign of **x** must be observed.

$$p_a = x_{a_1} - x_{a_2} = D - d_a$$
$$p_b = x_{b_1} - x_{b_2} = D - d_b$$
$$\begin{aligned} p_a &= x_{a_1} - x_{a_2} \\ &= D - d_a \\ &= D - (K - r_a) \\ &= (D - K) + r_a \\ &= C + r_a \\ C &= p - r. \end{aligned}$$

To complete the exercise, both parallax bar readings and parallax values are needed. This is achieved by using the principal points and their conjugates of stereopairs of photographs. See Figure 11.9.

$$p_{o_1} = x_{o_1} - x_{o_1'} = b'$$
$$p_{o_2} = x_{o_2} - x_{o_2'} = b$$
$$C_1 = b' - r_{o_1}$$
$$C_2 = b - r_{o_2}$$
$$C = \frac{C_1 + C_2}{2}.$$

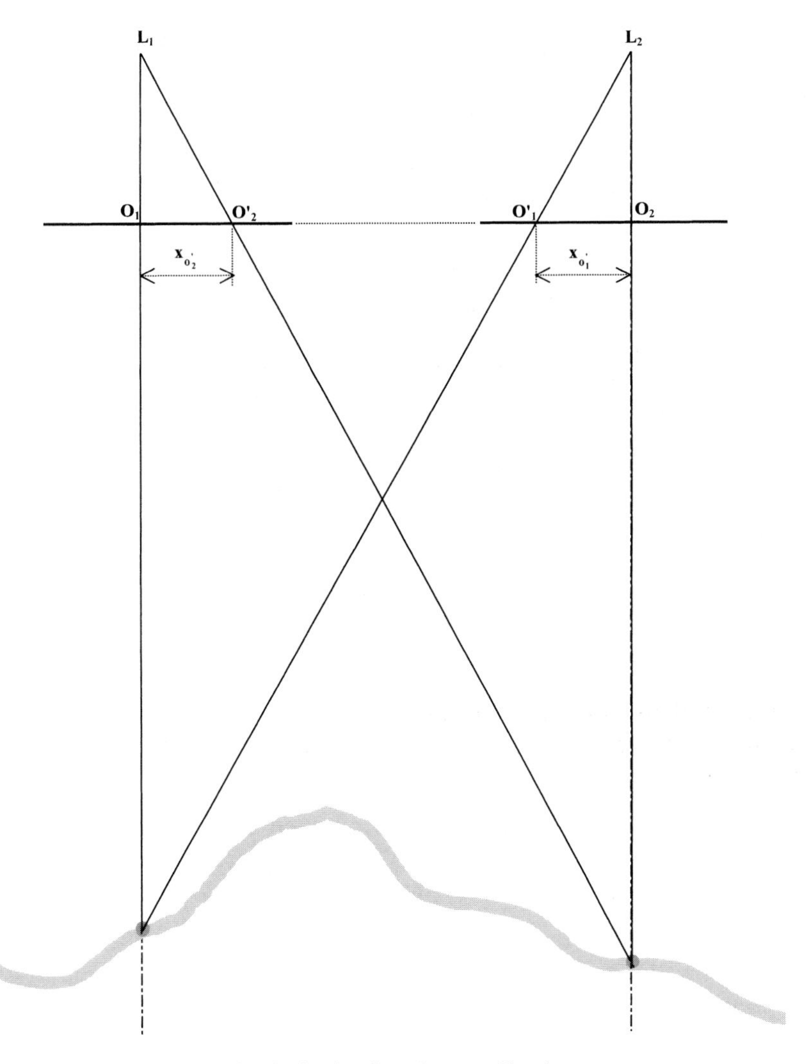

Fig. 11.9 Use of principal points for parallax bar constant

Now each parallax measurement can be computed from:

$$p = C + r$$

As **C** is an additive constant, it does not need to be determined if heights are being computed from parallax differences.

11.5 Planimetric Position from a Stereopair

The above methodologies only extract height from a pair of stereo photographs. For the extraction of horizontal position the compatible multi-image technique is called radial triangulation. The concept of radial triangulation also is the basis of the more advanced technique of *aerotriangulation* (*bundle/block adjustment*).

11.5.1 Radial Triangulation

The application of radial triangulation is based on the principal that, for a vertical photograph, the angle subtended at the principal point by two other image points is a true horizontal angle. This is the same as the horizontal angle that would be measured by a theodolite (or compass) situated on the terrain at the photograph's nadir point.

With a stereopair of photographs, angles to points in the overlap area can be measured from the two principal points using the camera base as the reference line. Their positions can then be plotted by intersection of the two angles. This is the same principle of intersection as used in field surveying – instead of using a theodolite on the ground, the angles are measured on the photographs. The whole process requires two operations:

(i) resection to determine the positions and orientation of the photographs (geo-referencing or registering);
(ii) intersection from two or more photographs to determine the position of new points.

As for the resection of a theodolite as given in Chapter 6, a minimum of three control points is required to define the orientation and scale of the produced plan. A simple methodology is given below.

(i) plot a grid at the required scale and the location of control points;
(ii) mark the principal and conjugate principle point on the first photograph;
(iii) overlay the photograph with tracing paper and plot the three control points and principal points;
(iv) draw radial lines from the principal point to each other point;
(v) orient the tracing on the plot sheet and mark the location of the principal point and the direction to the conjugate principal point;
(vi) from photograph, trace radial lines to detail in overlap area, transfer to plot sheet;
(vii) repeat on the overlapping photograph using the lines to the conjugate principal point and control points for orientation.

Steps one to five are the process of resection and steps six and seven are intersection. The radial triangulation of a stereopair can be extended to encompass a strip or block of photography for mapping large areas. However, it is a cumbersome method that is usually employed for small-scale plan production. A slight improvement in efficiency may be gained from using instruments such as the radial line plotter as shown in Figure 11.10. More

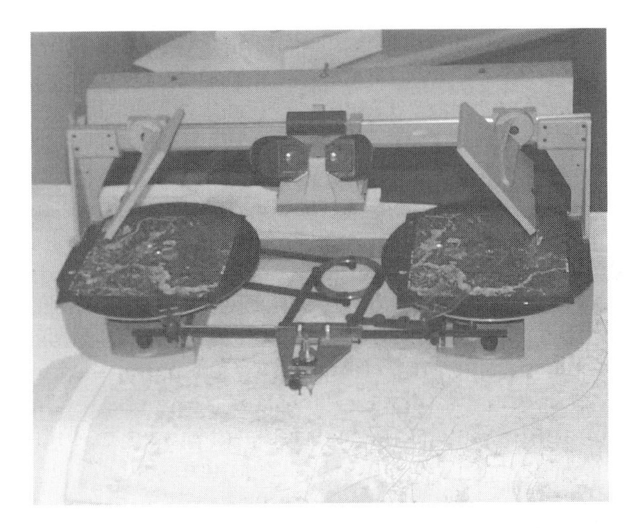

Fig. 11.10 Radial line plotter

efficient methods, based on analytical and digital principles, are commonly used and are described below.

11.5.2 Aerotriangulation

Aerotriangulation is the method of joining strips and blocks of photographs together through the application of radial triangulation to control, pass, and tie points. The use of control points has already been explained in section 11.5.1 but their use is minimised – control points are not required in every photograph. Instead, they appear in every 5 to 10 photographs. Each photograph is tied to the next by easily identifiable points called *pass points*, and strips are tied together by similarly identifiable *tie points*.

Aerotriangulation used to be a mechanical process. Modern techniques are analytical and have resulted in a method commonly called bundle adjustment or block adjustment. This process is a simultaneous computation along the lines of the forms outlined in section 11.6.

11.6 Analytical Methods

A critical point in the above methodologies is that by laying the photographs on a flat surface to perform the measurements, it is assumed that the photographs are vertical. Any deviation from this assumption generated by the camera axes not being truly vertical at the instant of exposure will have a significant effect on the accuracy of heights determined from parallax bar readings and positions determined by radial triangulation. Because of this assumption, these methodologies have limited application because of the inherent limitation of accuracy and so, for precise mapping purposes, more rigorous methodologies are required. These methodologies involve the determination of the actual tilts of the photographs and applying the results in the computation of height and position. These methodologies also require more sophisticated measurement environments such as the use *of analytical and digital stereoplotters*.

Along with the digital computer came the ability to digitise in three dimensions. Fundamental to the extraction of three-dimensional data from photographs is the geometric condition of *collinearity*. Simply put, this is the condition of a perspective projection (which is what a photograph is) where a point on the object, its image point, and the perspective centre of the photograph lie on the one straight line as is shown in Figure 11.11. This condition is described by the following equation:

$$
\begin{aligned}
x &= -c\frac{r_{11}(X_i - X_0) + r_{12}(Y_i - Y_0) + r_{13}(Z_i - Z_0)}{r_{31}(X_i - X_0) + r_{32}(Y_i - Y_0) + r_{33}(Z_i - Z_0)} \\
y &= -c\frac{r_{21}(X_i - X_0) + r_{22}(Y_i - Y_0) + r_{23}(Z_i - Z_0)}{r_{31}(X_i - X_0) + r_{32}(Y_i - Y_0) + r_{33}(Z_i - Z_0)}.
\end{aligned}
\qquad (11.4)
$$

This equation gives a direct link between points on an image and their three-dimensional, real world equivalent. However, as each equation represents a straight line, then conjugate image points need to be measured to compute the object point's three-dimensional position. This computation

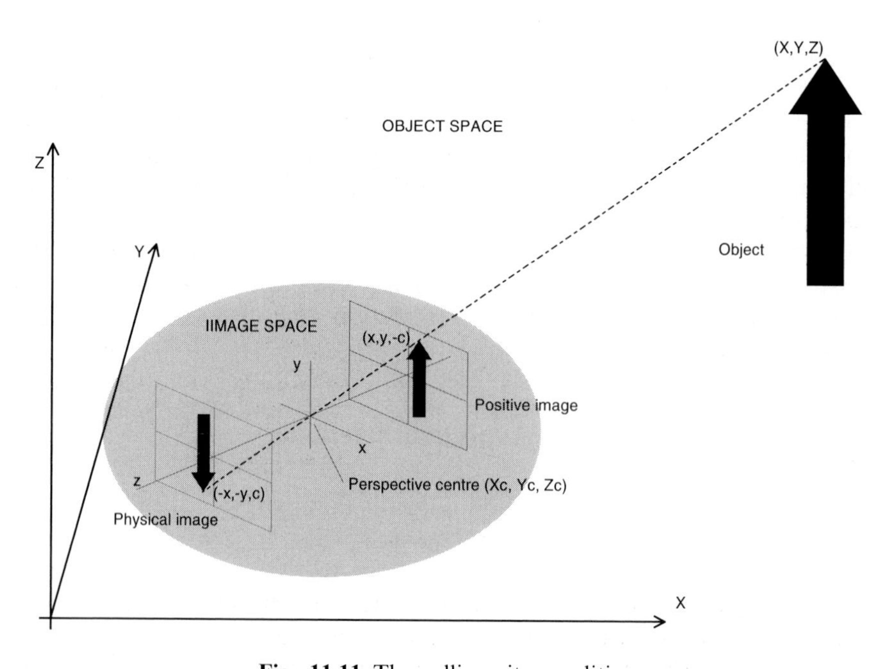

Fig. 11.11 The collinearity condition

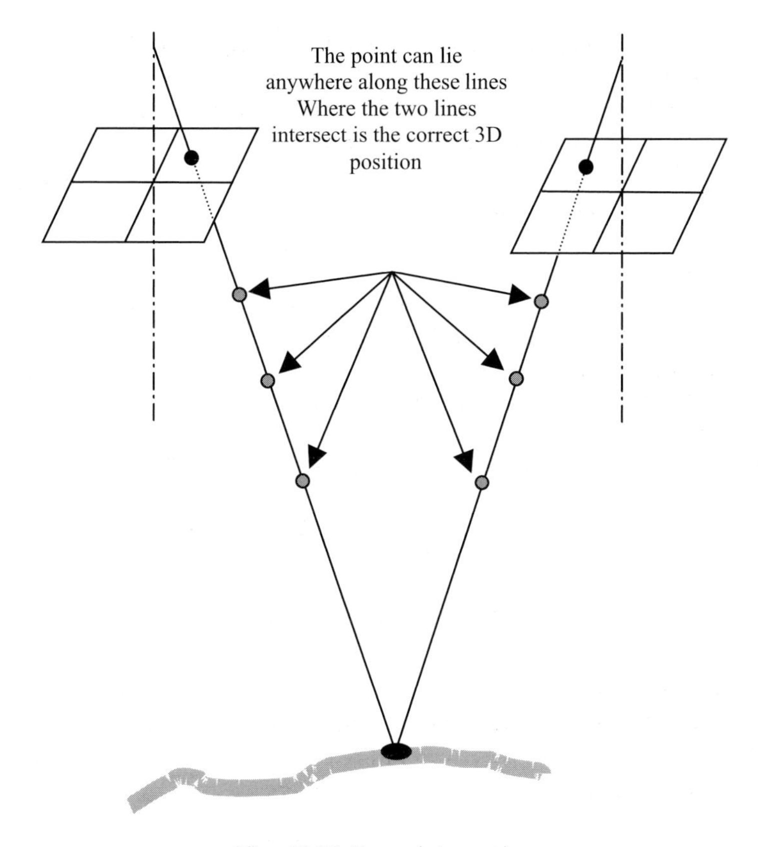

Fig. 11.12 Space intersection

is called *space intersection* and is achieved by computing $\mathbf{X_i}$, $\mathbf{Y_i}$, and $\mathbf{Z_i}$ from the above equation with observations from a stereopair of images. The concept is illustrated in Figure 11.12. The use of two photographs and the application of the collinearity condition may be considered as three-dimensional geo-referencing.

Before this computation can be made, various other parameters of the equation must be computed. First the image-space coordinates need to be determined through the process of *interior orientation* and second the *exterior orientation* parameters of the photographs are required. The former develops the relationship between the stereoplotter's measurement and the image-space coordinate systems; the latter the relationship between the image-space and object-space coordinates.

11.6.1 Stereoplotters

The stereoscope, parallax bar, and radial line plotters are relatively simple devices. More sophisticated devices that allow stereoscopic viewing of a stereopair and the simultaneous measurement of both horizontal and vertical positions as well implementation of the rigorous relationships between image-space and object-space coordinates are called stereoplotters. These machines have seen four generations of development:

(i) Projection – stereomodels formed through the use of image projectors (Figure 11.13).
(ii) Analogue – collinearity condition enforced and observations made by mechanical components on photographs (Figure 11.14).
(iii) Analytical – collinearity condition enforced mathematically and observations made electro-mechanically on photographs (Figure 11.15).
(iv) Digital – collinearity condition enforced mathematically and observations made digitally on digital images (Figure 11.16).

The sole purpose of these devices is to convert (two-dimensional) image-space coordinates to (three-dimensional) object-space coordinates. The processes used are varied and many and are beyond the scope of this book. The requirements for performing interior and exterior orientation exist in each of them. In most cases, projection and analogue stereoplotters have been superceded by analytical and digital instruments with digital instruments gaining wider application. The following sections will consider only the application of analytical and digital stereoplotters.

The space intersection problem is solved differently by analytical (and projection and analogue) and digital instruments. In the former, the operator

Fig. 11.13 Projection stereoplotter

Fig. 11.14 Analogue stereoplotter (from Leica)

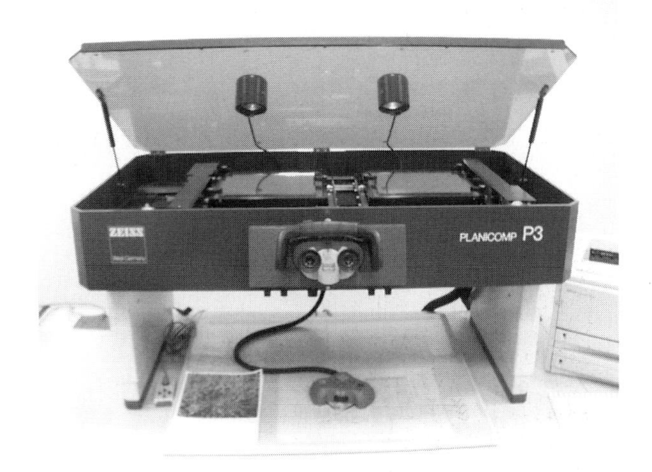

Fig. 11.15 Analytical stereoplotter

views the stereomodel following its interior and exterior orientation, and uses his perception of height through x-parallax to place the floating mark on the surface of the terrain – just as was done in the stereoscopic measurement of height. In digital systems, this process may be used (and is in low-end systems), but automated procedures that find the conjugate image points through *image processing* techniques are also employed.

11.6.2 Interior Orientation

The interior orientation relates the coordinates measured on the image to those of the object to be mapped. Before this can be done, two requirements

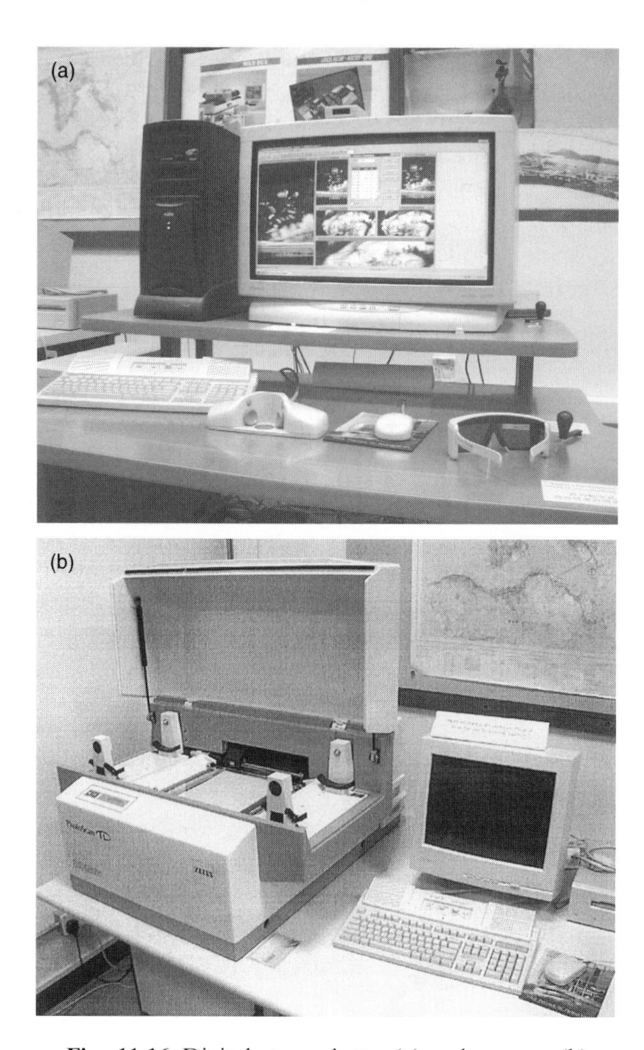

Fig. 11.16 Digital stereoplotter (a) and scanner (b)

must be met. First the relationship between the image coordinate and the measurement coordinate systems must be defined; and second the physical state of the image as it is being used compared to when it was originally obtained (such as film deformation) needs to be determined. This is the process of interior orientation. Regardless of whether the measurement system is analytical or digital, the basic principle is the same. For the first requirement, it is necessary to establish the location of the principal point and the orientation of the image coordinate axes. For the second, it is necessary to detect and correct for any systematic scale errors of the image. In the case of film-based images, scale errors may occur because of stretching or shrinkage (uniform or differential) of the film due to handling, processing, or storage. For digitally recorded images, differential scale changes may occur because of dissimilar pixel spacing or differences in pixel dimensions of the imaging device, or differences between the clock frequencies of the various imaging components (e.g., camera sampling rate and the frame grabber sampling rate). Ultimately, the shape of an object measured on the image needs to be that of the image when it was recorded. This is illustrated in Figure 11.17.

The fiducial marks (which have known coordinates in the image's coordinate system – see Figure 11.18) are measured. In the case of an analytical system, the observed coordinates (**X**, **Y**) are usually in millimetres; for a digital system the coordinates may be in either pixel coordinates or millimetres depending on whether the exact pixel dimensions are known. From these observations, a transformation between the observation and image coordinate systems is computed. There are several transformations used for this purpose. The most common being the two-dimensional conformal and affine transformations. Others include the two-dimensional rigid body and projective transformations. These transformations are discussed in detail in Chapter 3. Aerial photographs may have up to eight fiducial marks – one in each corner and one in the middle of each side. Images from a metric camera will have at least four corner fiducial marks.

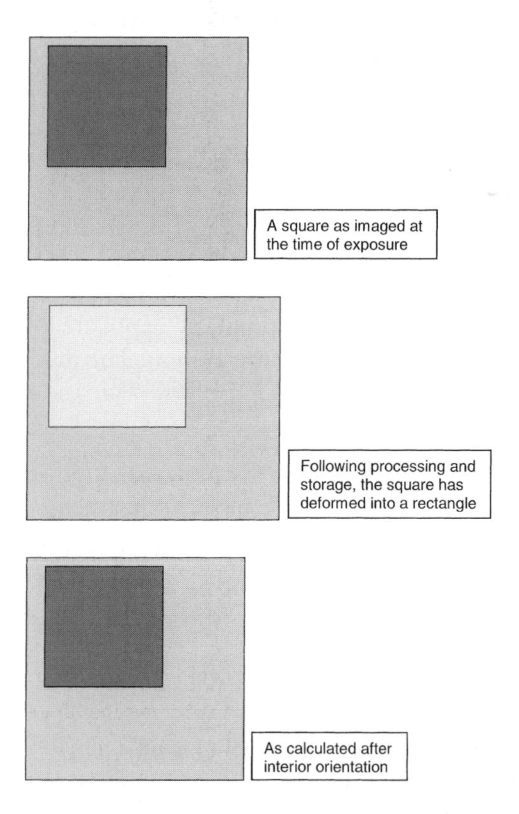

Fig. 11.17 The need for interior orientation

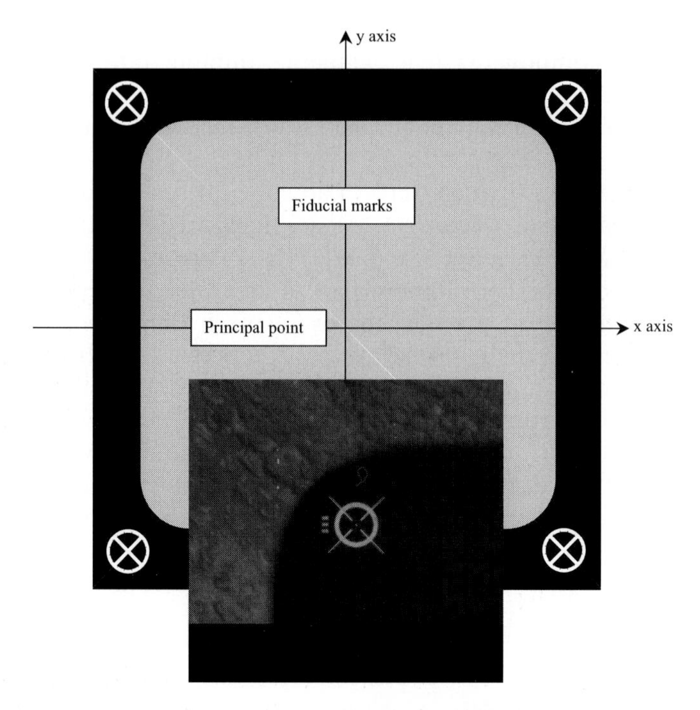

Fig. 11.18 Fiducial marks of an aerial photograph

11.6.3 Exterior Orientation

There are six parameters that describe the relationship between the photograph's and the object's coordinate systems. They are three angles, ω, ϕ, and κ, which describe the angular relationships, and $\mathbf{X}_0$, $\mathbf{Y}_0$, and $\mathbf{Z}_0$, which describe the location of the point of exposure of the images (perspective centres) in the object coordinate system. The process of determining these parameters is sometimes called *space resection* and can be achieved in one of two ways:

(i) relative orientation followed by absolute orientation, and
(ii) simultaneous solution by bundle/block adjustment.

These computations can be performed on individual stereopairs or on complete blocks of photography that cover the whole of the area to be mapped. In the latter case, the term aerotriangulation is often used to describe the procedure.

11.6.3.1 Relative Orientation

Relative orientation is based on the *coplanarity condition* and requires a stereopair for its determination. This condition states that the object point,

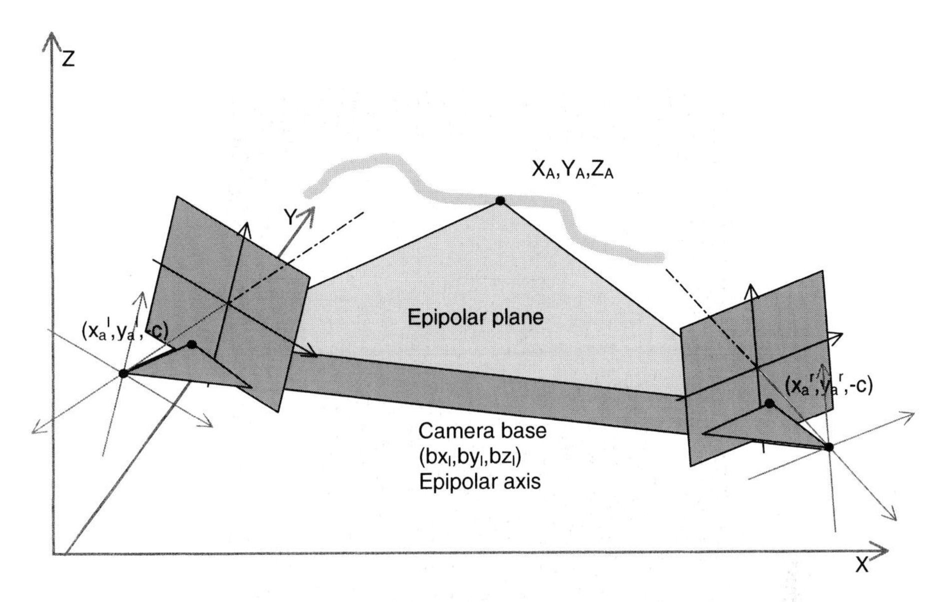

Fig. 11.19 The coplanarity condition

the perspective centres of the two images and the two image points lie on the same plane (see Figure 11.19). Once the condition has been fulfilled, through the observation of at least five common points on the two photographs (none of which is required to have known coordinate values), then a stereomodel has been established. At this time, however, the stereomodel is not accurately scaled nor are the three-dimensional model coordinates (the coordinates that can be observed in the stereomodel) directly related to the mapping coordinate system.

The common location of points used for relative orientation are the two principal points and either two or four other points distributed on either side of the principal point along the *y* axis of the images. In the case of using two additional points, they are often referred to as von Gruber points, and their locations are illustrated in Figure 11.20.

11.6.3.2 Absolute Orientation

This operation is a three-dimensional conformal transformation that converts the model coordinates obtained during relative orientation into correctly oriented mapping coordinates. To achieve this result, a minimum of two control points with horizontal coordinates and three control points with known heights must be visible in the stereomodel. With surveying technologies that readily produce three-dimensional coordinates with a single pointing, having a minimum of 3, three-dimensional control points is increasingly common. The methodology involves relating the image-space coordinates of the points of the relatively oriented stereopair to their known

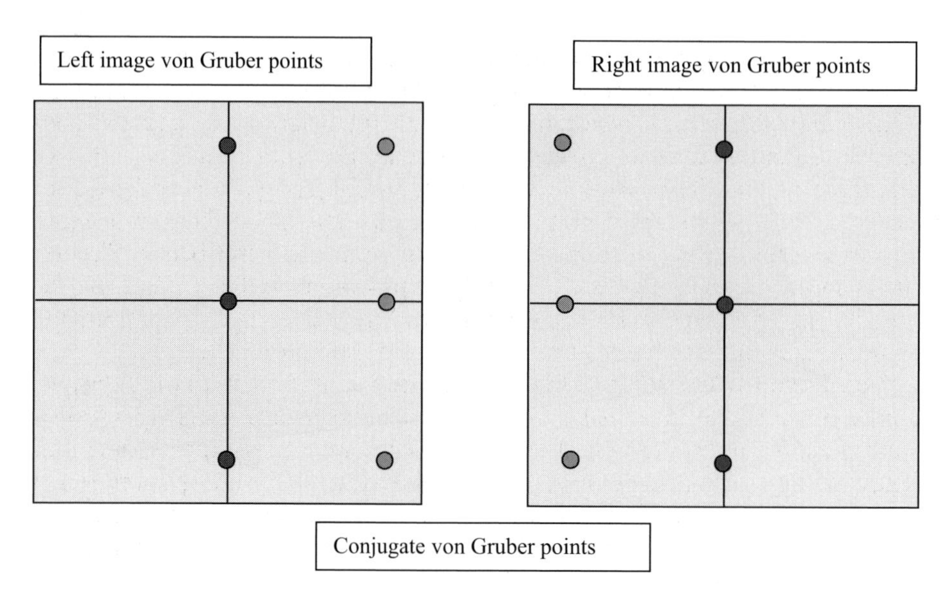

Fig. 11.20 von Gruber points

object-space coordinates. The transformation process is the same as that outlined in Chapter 3.

11.6.3.3 Simultaneous Orientation

An alternative to the two-step process of the relative/absolute orientation is that of the bundle/block adjustment. In this computation, the relationship between the photographs and the mapping coordinate system is computed in a single operation. The functional model of the bundle/block adjustment (the collinearity condition) can also be extended from that shown in equation (11.4) to include additional parameters that allow for the determination of uncorrected systematic errors. The systematic errors include lens distortion, film deformation, atmospheric refraction, earth's curvature, principal distance, and principal point location that may not have been rigorously determined (through calibration) or corrected for at the time the image point observations were made.

This process may be applied on a stereopair-by-stereopair basis or upon the entire block of photographs. There are several differences between the approaches, and the adoption of one over the other relates to the scope of the project.

For the former, control points – a minimum of 3, three-dimensional points – are required to be present in the overlap area of each stereopair. Observation of the image-space coordinates (following interior orientation) of each of the control points visible in the overlap area will allow the exterior orientation parameters to be computed for both photographs.

If three overlapping photographs are to be used, however, then there will be two orientations to be done with the middle photograph being used twice. Observation error may result in the middle photograph having different orientation parameters (the degree of difference depending on the quality of the observations). This leads to a difference in object-space coordinates of features that may be found in both stereopairs – similar to the mono-photo process of separately registering overlapping photos as described in Chapter 10. Also, if a strip or block of photographs were to be used, then the coordinates of a large number of control points will need to be determined by field surveying methods as described in Chapters 6 and 7.

To overcome these limitations, the exterior orientation of all photographs can be computed at once using the concept of analytical aerotriangulation – the bundle/block adjustment. The prime advantage of this approach is that following the computation a single set of exterior orientation parameters is computed for each photograph and problems such as those described above are minimised. The entire block of photographs now exists in a single homogenous coordinate system.

11.6.4 Three-Dimensional Digitising

Following exterior orientation, the full relationship between the stereomodel and the mapping coordinate system is known. Any point measured in the stereomodel can be directly related to the features that can be seen on the imaged terrain. The parameters of the absolute orientation can then be used in the collinearity equations. The only unknown values from further observations are then $\mathbf{X_i}$, $\mathbf{Y_i}$, and $\mathbf{Z_i}$, the coordinates of the observed points. These points can be either discrete point features, such as utility covers or sign posts, or spot heights on natural or artificial surfaces; the beginning or end or intermediate points of curvilinear features, such as roads, rivers, and contour lines; the junction points of two or more curvilinear features; or the corners of polygonal features, such as building corners and shorelines.

The process of measuring these detail points and computing their three-dimensional coordinates is often referred to as *feature extraction.* The process of feature extraction is simply that of space intersection. In both analytical and digital stereoplotters, the operator views the stereomodel and digitises in three dimensions those points that constitute features to be captured. $\mathbf{x}$ and $\mathbf{y}$ image-space coordinates of the conjugate image points are converted through the use of the collinearity equations into $\mathbf{XYZ}$ object-space coordinates along with the appropriate attribute data that describes the feature being collected.

When operating in a stereoscopic environment, it is important that the floating mark be kept on the surface of the feature; otherwise the three measured coordinates will be incorrect. This is a common mistake made by students starting out in a three-dimensional environment. More concentration is placed on controlling the movement of the floating mark over the planimetric position of the stereomodel and the height coordinate is neglected.

The digitised features are often stored directly as point, line, and polygon features in mapping system or CAD files. The data is collected by means of an interface to general-purpose software such as MicroStation and AutoCAD, to specialist mapping software such as KORK, or directly to GIS software such as Arc/Info or CARIS.

11.6.5 Accuracy Considerations

With such complex relationships between the image point and object point, it is difficult to give simple indicators of accuracy and precision for data obtained stereoscopically. Analysis of the basic stereoscopic measurement model yields the following relationships:

$$\sigma_{X,Y} = \frac{Z}{f}\sigma_{x,y}$$
$$\sigma_Z = \frac{Z}{f}\frac{Z}{B}\sigma_p. \tag{11.5}$$

For good quality ordinary photogrammetry:

$$2,000 < \frac{Z}{f} < 100,000$$
$$1 < \frac{Z}{b} < 4 \tag{11.6}$$
$$5\,\mu m < \sigma_{x,y} < 20\,\mu m$$
$$2\,\mu m < \sigma_p < 10\,\mu m$$

Typically, it is recognised that the vertical precision of data is approximately half that of the horizontal precision. More accurate estimation of data precision can be obtained from the results of aerotriangulation computations.

11.7 Digital Methodologies

In recent years, considerable advances have been made in the area of automated or semi-automated processing of images. These advances are all due to the application of the technology of *image processing* to solving photogrammetric problems. Image processing has been applied to almost every aspect of photogrammetry from capturing the imagery (by scanning or digital cameras) to processing the imagery in digital photogrammetric workstations (DPWs). Image capturing technology is covered in Chapter 9, and details of image processing are given in Chapter 12.

In analytical photogrammetry, the previously described methods rely on an operator to view the photographs and place the floating mark in the necessary positions (over fiducial marks or on the ground.) The resulting data essentially consisted of vector-based information (point, line, and surface) and digital elevation models. To perform this operation over an extended period of time and with high accuracy and precision requires extensive training. Computer technology is now being used to automate this process. The task of placing the floating mark is being replaced by a class of image processing techniques called *image matching* that search and find the conjugate points on a pair of digital images, and from this information compute a point's **X**, **Y**, and **Z** coordinates. Such processes automate and speed up the extraction of two- and three-dimensional information from a stereopair and is the domain of digital photogrammetry. An additional advantage of digital processing is that products can also be of the raster variety – digital images rather than vector data.

Before the extraction of three-dimensional information can begin, the previously mentioned processes of interior and exterior orientation must also be undertaken. Image matching also plays a role here in finding and measuring the locations of fiducial marks and control points.

11.7.1 Image Matching

In general, image matching (also called *image correlation*) is based on either examining and matching the grey levels of small portions (image patches) of each photograph of a stereopair, or matching an image patch of one photo with an image template. The matching may be on a pixel-by-pixel basis (area-based matching) or by examining and matching individual features of the image patches (feature-based matching).

Regardless of the methodology used, the most important applications of image matching are: the automatic interior and exterior orientation of an image; the automatic creation of DTMs from stereo photography; and the extraction of features in three dimensions such as roads, buildings, and natural boundaries. With analogue and analytical systems, each of these operations was controlled by the operator, but in a digital environment the role of the operator has changed to that of a manager and editor of automatically generated information. An important issue to keep in mind here is the reliance on traditional aerial image geometry for the successful extraction of terrain. Experience has shown that using some DPWs with oblique or convergent imagery has not been as successful.

Of prime importance for all image matching operations is that the image must have sufficient *contrast* and *texture* to allow the algorithms to sufficiently operate. Other limitations on the successful application of image matching is that of *dead areas* caused by shadows and relief displacement and also discontinuities caused by relief displacement. As a result, an important feature of DPW software is that of stereoscopic terrain editing functions.

It has been found that image editing is a significant task in the automatic generation of DTMs as the image matching systems are far from infallible.

There are many approaches to image matching so only the basic concepts are covered here. This section introduces only the two basic matching techniques mentioned above. As this field of research is in a state of rapid change any further discussion will quickly be out-dated.

To improve the speed of operation and to ensure that stereoscopic vision is maintained, the images of a stereopair are first re-sampled along epipolar lines. Thus conjugate image points will appear very close to the same row on each of the two images.

11.7.1.1 Area-Based Matching

For this approach, the locations of conjugate patches are found by comparing the grey levels of one image (reference image) or image patch (the reference patch) with the same sized patches of the other image (the search image or search patch). The reference and search patches are said to match when the difference between the grey scales (digital numbers) of each pixel in the image patches is a minimum according to some criteria. A commonly used simple criteria to assess the degree of agreement between the two patches is the cross-*correlation coefficient*:

$$K = \frac{\Sigma(g_1 - \bar{g}_1)(g_2 - \bar{g}_2)}{\sqrt{\Sigma(g_1 - \bar{g}_1)^2 \cdot \Sigma(g_2 - \bar{g}_2)^2}}$$

where $\mathbf{g}_1$ and $\mathbf{g}_2$ are the grey levels for each corresponding pixel in patches 1 and 2 and $\bar{\mathbf{g}}_1$ and $\bar{\mathbf{g}}_2$ are the mean grey levels of the patches. The reference patch image is moved over the search image to find the maximum value of K. The success of this method of area-based matching is limited to images that have similar geometric and radiometric properties such as those from overlapping aerial mapping images of uniform terrain.

More sophisticated methods, such as *least-squares matching,* have been developed to allow for a wider range of geometric and radiometric differences between the reference and search patches. Imagery taken with different cameras, with convergent camera axes, or on steeply sloping terrain means that the apparent shape and size of an object to be matched on one image may be quite different to that on another image. Least-squares matching recognises this and can adapt to such changes much better than can the cross-correlation method.

Depending on the nature of the image matching task, the reference patch may be a specific pattern such as the image of a fiducial mark or some other pre-defined object or a portion of one of the stereopair images. In the former application the reference patch is known as a template, and the operation of finding the template on the searched image is called template matching.

The results of area-based matching can be very precise. The image coordinates of matched objects have been reported to have a precision of better than 0.1 pixel. On an image with pixel size of 20 μm (commonly used for aerial photogrammetric tasks), this represents a precision comparable with the best manual observations on an analytical stereoplotter. Template matching can achieve precision of up to 0.01 pixel.

11.7.1.2 Feature-Based Matching

Whereas area-based matching examines portions of an image on a pixel-by-pixel basis, feature-based matching takes an overview approach. The aim of this method is to find distinct features on the reference image and then find them on the search image. In simple terms, a feature is merely a distinct change in image brightness compared to the surrounding pixels.

Fundamental to the application of feature-based matching are image pyramids. An image pyramid is a series of images each at half the resolution of the previous image. At the bottom of the pyramid is the image of full resolution. The next level image is obtained by combining each 2×2 block of pixels to form an image of half the resolution. The next level image combines a 4×4 block of pixels (of the original image) and so on depending on the size of the original image. Some digital systems limit the size of the image pyramid to, for example, eight levels. This gives a reduction of 1/256 of the original image. The image pyramid is also useful in providing image "zoom" functions. The amalgamation of each block of pixels is achieved by one of the re-sampling methods described in Chapter 12. A simple image pyramid is shown in Figure 11.21.

The purpose of building the image pyramid is to speed up the feature detection process. It has been found that distinct features can be seen on lower resolution images just as effectively as on higher resolution images. However, processing lower resolution images is much more efficient than higher resolution ones because of the fewer number of pixels to be examined. Following the identification of prominent features on an image, its location is traced through the image pyramid at increasingly higher levels of precision as more detail is seen at higher image resolutions. Once a series of features have been located on the lower resolution image, only those areas on the increasingly higher resolution images need be examined to accurately define the location of the feature. Finally, precise image measurement is achieved by area-based matching techniques.

11.7.2 Automatic Feature Extraction

This topic was also dealt with in Chapter 10. The challenge here is to accurately identify the same features on a stereopair of photographs. This image matching methodology, as mentioned in the previous sections, must

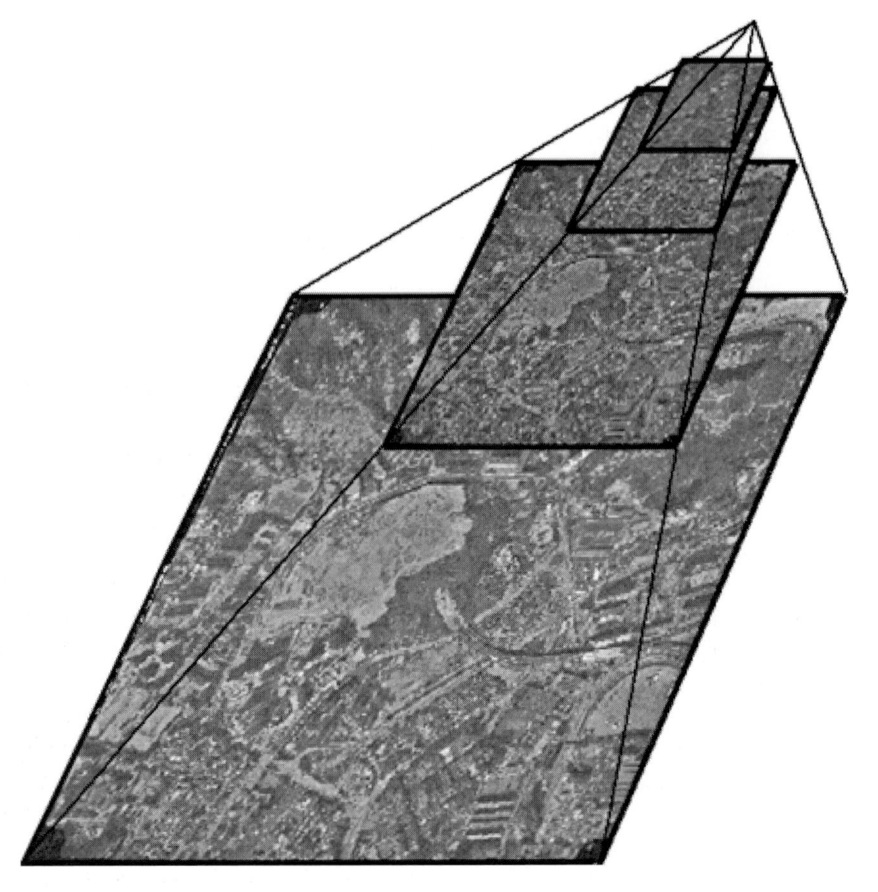

Fig. 11.21 An image pyramid

be employed in order to extract three-dimensional features. In effect the process involves replicating the human operator's ability to interpret the imagery and to recognise conjugate image points in the pair of images.

Automation of this task has been achieved for imagery containing continuous surfaces such as natural terrain and artificial objects. However, with discontinuous features, such as cliff edges, buildings, and multi-level objects such as roadway overpasses, the problem of relief displacement causes considerable difficulty to automatically locating the conjugate image points. When such a feature is found in one image, without knowledge of the height of the object, it is difficult to predict where it will be found in the other image. Add to this the problem of seeing the vertical surface of such an object in one image and not the other, makes the problem significant. It is solving this problem – something the human operator does very well – that researchers are concentrating on at the present time.

12 Thematic Information from Digital Images

Qiming Zhou

Digital images, particularly those from remote sensing technology, have become an important source of spatial information. In modern Geographical Information Systems (GIS), digital remotely sensed images are widely recognised as one of the most practical means of spatial information updating, especially in real-time applications. In most of today's applications, however, the remotely sensed data may only be used to their greatest potentials if they can be correctly interpreted, classified, and presented in the same way as other terrestrial spatial information, such as thematic maps.

This chapter discusses the methodologies and techniques of extracting thematic information from digital images. As an introduction, the nature of digital images and the characteristics of objects on earth in relation to image interpretation are described. Then the discussion focuses on the techniques of image interpretation and auto-classification using black-and-white (or single band) images or multispectral images. Methods and techniques used for integrating digital images with spatial information systems are also revealed in this chapter. For the purpose of this chapter and ease of discussion, the digital images referred to in this chapter include only those taken from passive remote sensing.

12.1 Factors to be Considered for Thematic Information Extraction

Remotely sensed images are presented in a way similar to a normal photograph, except that digital numbers are used to represent brightness values. To extract thematic information from these digital numbers, three basic factors must be considered, namely, spectral, spatial, and temporal characteristics. The spectral characteristics refer to the nature of electromagnetic radiation that is emitted or reflected from the earth's surface and the capability of the sensor to detect it. The spatial characteristics describe the size of the earth's objects to be detected in comparison with the spatial resolution of the sensors. Given that we live in a changing world, the temporal characteristics must be considered while monitoring our environment, not only about the nature of phenomena, but also about the capability of the sensor.

12.1.1 Spectral Characteristics

Human eyes are a form of remote sensing detectors. They react to light that comes from or is reflected from things around us. The light that human eyes detect, however, is just a small portion of a continuous spectrum of energy called the *electromagnetic spectrum.*

The electromagnetic spectrum is made up of energy that is known as *electromagnetic radiation* (EMR) because the pulses or waves can be measured both electrically and magnetically. Specific names are used to describe the different wavelengths that make up the electromagnetic spectrum (Figure 12.1). The light that human eyes can detect is called visible light because it is visible to the naked eye. Indeed, there is nothing particular about the visible portion of the electromagnetic spectrum other than the fact that humans can see it.

Most remote sensing is based on the detection of EMR, whether the detector is a human eye, a camera, or a scanner on a satellite (an electronic "camera"). What can be learned from remote sensing depends on what type of radiation is detected and how it interacts with the surface of the earth (or any other surface that is being sensed). Human vision is particularly important because any information that is collected generally is interpreted visually from a printed image or on a computer screen.

12.1.1.1 The Electromagnetic Radiation from Earth Materials

All matter in the universe that is warmer than 0°K (or approximately −273°C) emits electromagnetic energy. Molecular motion stops at 0°K – the coldest possible temperature, which is also known as *absolute zero.* All objects in every day life are warmer than absolute zero.

The wavelength of EMR ranges from very short (nanometres) to very long (kilometres) as shown in Figure 12.1. The amount and type of energy that is emitted depends on the temperature of the object. Very cold objects only emit energy with very long wavelengths, while warm objects emit both long and

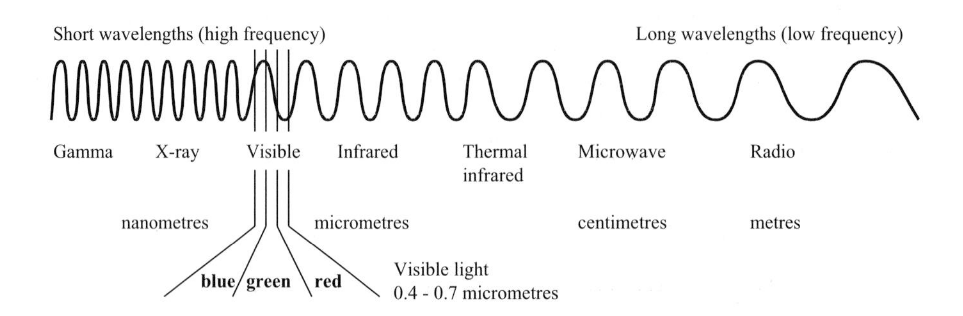

Fig. 12.1 The range of wavelengths known as the electromagnetic spectrum. Our eyes can only see a small part of this energy known as visible light

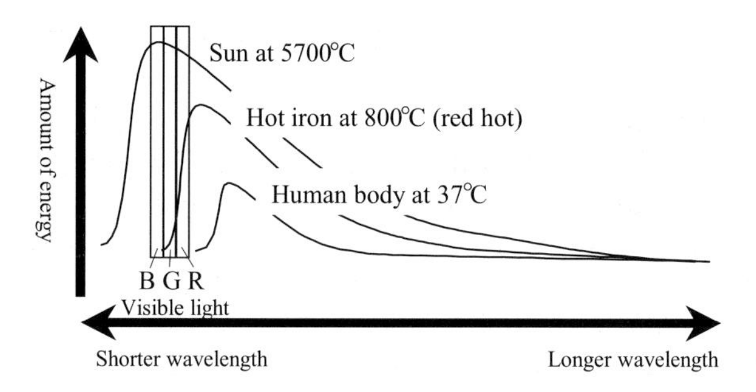

Fig. 12.2 The wavelength and amount of energy that is emitted from the sun (top curve), hot iron (such as a glowing stove element, middle curve), and the human body (bottom curve)

short wavelengths. In addition to emitting a wider range of wavelengths, a warmer object emits more energy than a cold object. This higher energy level is caused not only by an increase in the amount of EMR being emitted, but also because shorter wavelength EMR contains more energy.

The term *blackbody* refers to a surface that absorbs all incoming energy (and hence, looks black) and re-emits all that energy. To quote Lillesand and Kiefer [1994, pp. 7] "a blackbody is a hypothetical, ideal radiator that totally absorbs and re-emits all energy incident upon it. Actual objects only approach this ideal." The wavelength at the peak of the blackbody curve is the wavelength of maximum emittance, which is directly related to the temperature of the surface.

A surface needs to be very hot before it emits energy at such short wavelengths. Figure 12.2 shows a diagram of the amount of energy and the wavelengths of energy emitted from the sun, a hot stove element, and the human body. The light from the sun is slightly yellow because there is more green and red than blue light being emitted. In addition, some of the blue light that is emitted by the sun is scattered by the atmosphere, further reducing the amount of blue and making the sky appear blue. A hot stove element looks red because it is not hot enough to emit blue and green light, but only red light and energy with longer wavelengths than we can see (infrared and longer). The energy that is emitted from the human body cannot be seen at all by human eyes because of the energy is at wavelengths that are much longer than the eye can see.

A surface that emits energy close to that of a blackbody is called a *greybody*, while a surface that emits energy close to a blackbody at some wavelengths but not at others is called a *selective radiator*. Most of earth materials are selective radiators (Figure 12.3).

12.1.1.2 Spectral Characteristics of Sensors

One important consideration in selecting the wavelength range in which a remote sensor will detect is the *atmospheric transmittance*. Earth's

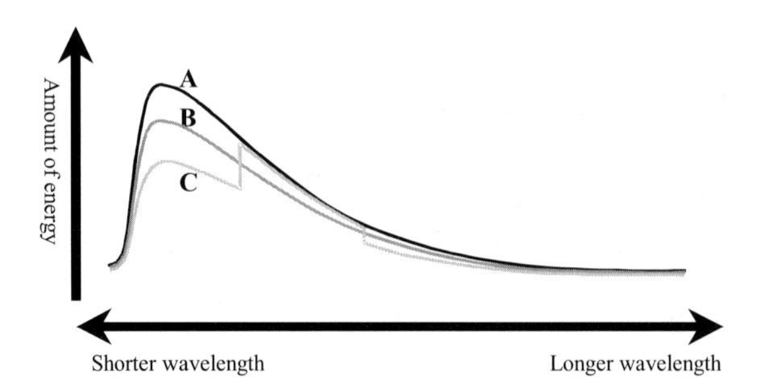

Fig. 12.3 Curves A, B and C show the energy emitted from three surfaces with different emission characteristics: (A) a blackbody, (B) a greybody, and (C) a selective radiator

atmosphere itself selectively scatters and absorbs energy in a certain spectral range, allowing the rest of the solar energy to transmit through it. Areas of the spectrum where specific wavelengths can pass relatively unimpeded through the atmosphere are called *transmission bands*, or *atmospheric windows*, whereas *absorption bands*, or *atmospheric blinds*, define those areas where specific wavelengths are totally or partially blocked. For a remote sensor that is capable of 'seeing' objects on the ground, the detectors must use the transmission bands (Figure 12.4).

The spectral responses that a remote sensor can 'see' are dependent upon the spectral bands that the sensor detects. In remote sensing, the spectral range is usually composed of a number of spectral bands (which fall within the 'atmospheric windows'), ranging from a single band image (panchromatic image) to several hundred images (hyper spectral image). Usually, the term 'multispectral' is applied to the images that are composed of several spectral bands. The spectral characteristics of commonly used space-borne sensors; their spectral bands and primary use are listed in Table 12.1.

12.1.1.3 Spectral Signatures of Some Earth Materials

For passive remote sensing, the 'light' that a sensor detects is mainly reflected sunlight, which has an energy distribution over the entire spectrum, although some sensors do have the ability to detect energy emitted from Earth's surface itself (e.g., thermal infrared). In theory, the reflected sunlight from different kinds of materials found on Earth varies with the material. The spectrum of the reflection for a certain material is often unique; therefore it is called the *spectral signature* of the material. In remote sensing, it is fundamental to investigate the spectral signature before a correct image interpretation may be achieved.

There is a huge variety of materials on Earth's surface, therefore recording their spectral signatures (also known as a *spectral library*) requires substantial investments of money and time. For years, efforts have been made to establish such spectral libraries and some have already become available,

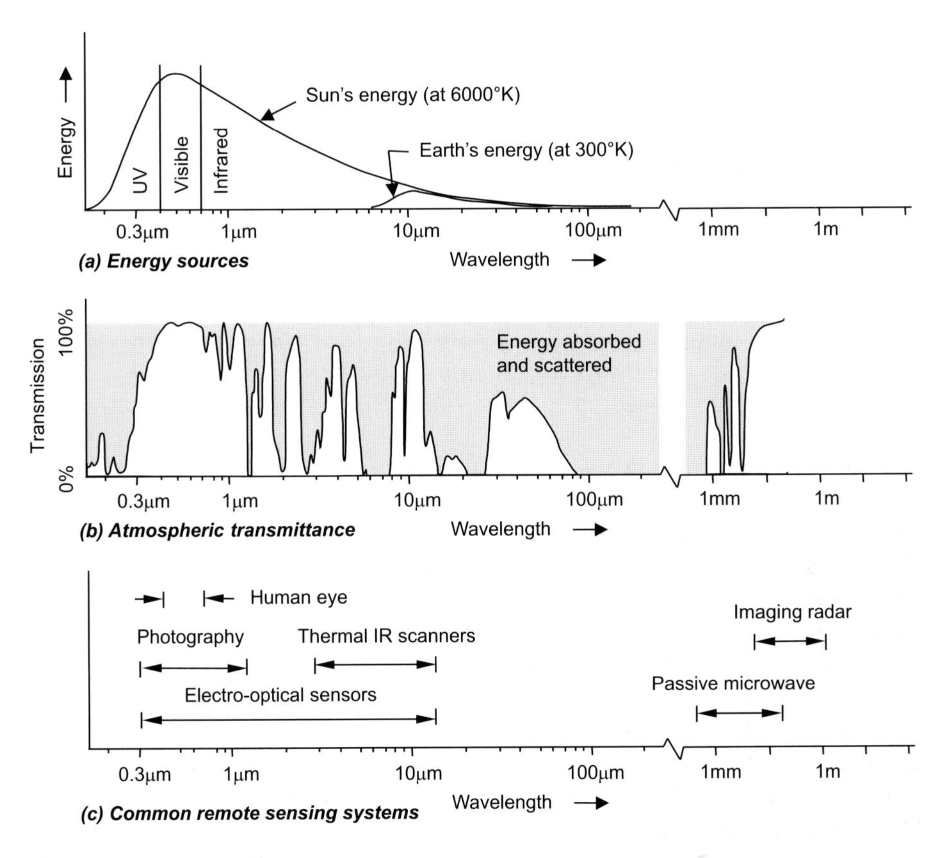

Fig. 12.4 Spectral characteristics of energy sources, atmospheric effects and remote sensing systems [after Lillesand and Kiefer, 1994]

often associated with remote sensing image processing software packages. The spectral signatures of some of Earth's typical materials, such as vegetation, soil, and water, are now fairly commonly known (Figure 12.5).

For information extraction and image interpretation, the selection of appropriate bands of multispectral images for the objectives of the application is a crucial task. The comparison of multispectral bands that present the most distinct difference between the cover types of interest gives the most promising hope for correct interpretation and classification, whereas difficulties are often experienced in separating cover types with image bands that record spectral regions where the cover types present a similar response.

12.1.2 Spatial Characteristics

Another important factor for information extraction from digital images is the spatial extent of the objects to be interpreted and the sensor's spatial resolution. In theory, if the objects are smaller than the image resolution,

Table 12.1 The spectral characteristics of some currently operational space-borne remote sensors

Satellite	Sensor/ Image	No. of Bands	Band	Spectral Range (μm)	Primary Use
Landsat	MSS	4	1	0.5–0.6	Culture features, clear water penetration
			2	0.6–0.7	Vegetation/soil discrimination
			3	0.7–0.8	Delineating water bodies, geology
			4	0.8–1.1	Delineating water bodies, vegetation vigour and biomass
	TM	7	1	0.45–0.52	Coastal water mapping, soil/vegetation discrimination, forest type mapping, and cultural features
			2	0.52–0.60	Vegetation discrimination and vigour assessment, and cultural features
			3	0.63–0.69	Plant species differentiation, and cultural features
			4	0.76–0.90	Vegetation types, vigour, and biomass content, delineating water bodies, and soil moisture
			5	1.55–1.75	Vegetation and soil moisture content, differentiation of snow from clouds
			6	10.4–12.5	Vegetation stress analysis, soil moisture discrimination, and thermal mapping applications
			7	2.08–2.35	Discrimination of mineral and rock types, and vegetation moisture content
SPOT	PAN	1	1	0.51–0.73	General mapping, DTM generation
	XS	3	1	0.50–0.59	Cultural features, clear water penetration
			2	0.61–0.68	Vegetation/soil discrimination and plant species differentiation
			3	0.79–0.89	Delineating water bodies, vegetation types vigour and biomass
NOAA	AVHRR	5	1	0.58–0.68	Daytime cloud and surface mapping, snow and ice extent
			2	0.725–1.1	Surface water delineation, snow and ice extent
			3	3.55–3.93	Detecting hot targets (e.g., forest fires), night-time cloud mapping
			4	10.3–11.3	Determining cloud and surface temperatures, day or night cloud mapping
			5	11.5–12.5	Determining cloud and surface temperatures, day or night cloud mapping, water vapour correction

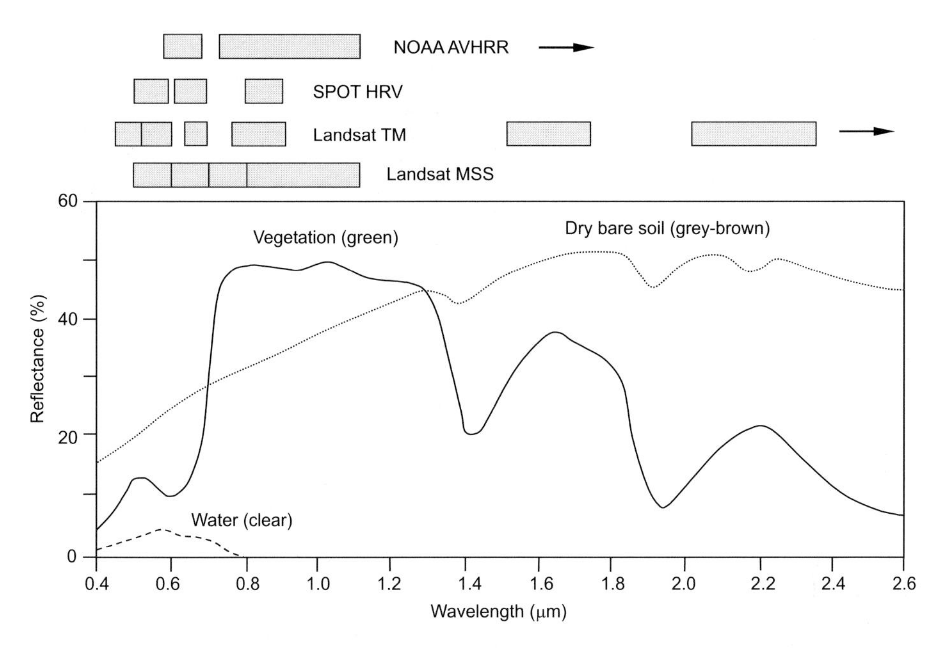

Fig. 12.5 Typical spectral reflectance curves of common earth surface materials in the visible and near-to-mid infrared ranges. The positions of spectral bands for some remote sensors are also indicated [after Richards, 1993]

the objects cannot be reliably perceived and correctly interpreted. In reality, however, some small objects may be visible on images with a lower resolution, provided that there is enough contrast between the objects and their background. In Figure 12.6, the linear features (left of the image) in an ancient lake are clearly visible indicating the road crossing the area. Since there is a high contrast between the road and its surrounding background, it is shown on the image in spite of its width being far less than the 82 m resolution of the MSS imagery.

According to sampling theory, the spatial resolution of the digital image must be at least half that of the smallest object that is of interest, so that the shape of the object can be reliably presented on the image. This constitutes

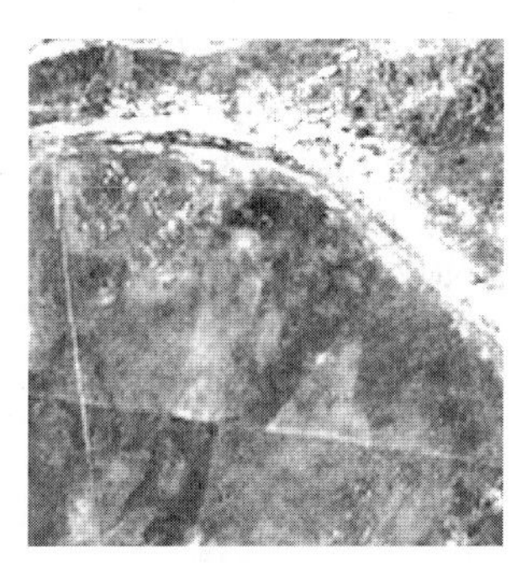

Fig. 12.6 Landsat MSS image of an ancient dry lake near Pooncarie, western New South Wales, Australia

Table 12.2 Examples of spatial resolution of some commonly used remotely sensed imagery

Platform	Sensor/ Image	Spatial resolution	Interpretation features	Primary applications
Aircraft	Digital airphoto	1–2 m	Control points on cultural features	Photogrammetry and mapping, urban management
SPOT	PAN	10 m	Houses and streets	Urban planning
	XS	20 m	Crop fields, water bodies, urban areas	Regional planning, agriculture, land use change
Landsat	TM	30 m	Crop fields, water bodies, urban areas	Regional planning, agriculture, land use change
	MSS	82 m	Landforms, forest, pasture and agriculture areas	Environment and pasture, rangeland management
NOAA	AVHRR	1.1 km	Regional landforms, coastline	Regional monitoring, coastline and oceanography
GOES	GOES	2.5–5 km	Clouds, coastline	Weather forecast, oceanography, global change

the bottom-line limitation of the digital image concerned. Table 12.2 outlines the spatial resolution of commonly used remotely sensed data and their interpretation features and primary applications related to their spatial resolution.

One of the greatest challenges in remote sensing is the development of techniques and methodologies that deal with so-called '*mixed pixels*', which are pixels covering different types of thematic information [Schowengerdt, 1997].

12.1.3 Temporal Characteristics

Since we are living in a changing world, frequent regular monitoring of our environment is one of the major application areas for remote sensing. Multitemporal remotely sensed images make this possible.

For thematic information extraction, the temporal factors that may influence the interpretation process and potential applications include, for example, acquisition date and time, frequency of coverage, and history of coverage.

12.1.3.1 Acquisition Date and Time

Date and time, in addition to geographical position, are often the first criteria applied to the acquisition of remotely sensed data. For many applications, it is vital to acquire simultaneous imagery (or near simultaneous) to be matched with the 'ground-truthing' information. Appropriate timing for image acquisition can also help interpretation. For example, the effect of shading and shadow, which can vary on different acquisition dates (e.g., winter or summer), may provide important information for the interpretation of topography. Day and night images acquired by some sensors may be used to distinguish different types of surface materials.

12.1.3.2 Frequency of Coverage

The frequency of coverage of a particular type of remote sensor determines how often we may use the derived images to monitor a given area (Table 12.3). For most of today's commonly available satellite data, which are basically orthographic in nature, the re-visit frequency is determined by the satellite orbital characteristics. A satellite designed for Earth resource monitoring usually takes a *sun-synchronous* orbit, meaning that the satellite passes over all places on the earth having the same latitude at approximately the same local time. However, some satellites (e.g. SPOT) have the ability to take an 'order' to acquire 'side-looking' images, thus providing more frequent coverage than using only the orthographic images.

For some applications, such as flood and bush fire monitoring, the re-visit period is crucial. These events have a critical demand for timely coverage and situations may change rapidly. Generally speaking, for a given sensor, the re-visit period and the spatial resolution are negatively related. This is because the lower resolution imagery often covers a larger area so that the same area would be more frequently covered again. Given this fact, it is conceivable that most of today's large area remote sensing applications

Table 12.3 Temporal characteristics of some commonly used satellite data and their area coverage

Satellite	Sensor/ Image	Daylight crossing local time at the equator	Re-visit period at the equator	Pixel resolution	Swath width
SPOT	HRV PAN	10:30 am	26 days up to 7 passes/26 days	10 m	60 km
	HRV XS			20 m	
Landsat	MSS TM	9:45 am	16 days	82 m 30 m	185 km
NOAA	AVHRR	Even No.: 7:30 am/pm Odd No.: 2:30 am/pm	2 passes/day	1.1 km	2800 km
GOES	GOES	Geo-stationary		2.5–5 km	Full-disk view

that rely on land monitoring would most likely use lower resolution data such as NOAA AVHRR.

12.1.3.3 History of Coverage

Since the Landsat satellite was first launched, we have accumulated nearly 30 years of regularly collected satellite-borne imageries. This does not include the even longer history of collecting aerial photographs and data from meteorological satellites. The historical collection of remotely sensed imagery gives us an opportunity to study long-term change in and human impact on many aspects of our living environment.

12.2 Feature Identification and Image Interpretation

Digital remotely sensed data are becoming one of the major data sources for database updating. We must go through a classification step, however, to extract thematic information from the digital images. Correct feature identification and image interpretation in a given application context, therefore, are crucial to the successful utilisation of remotely sensed information.

12.2.1 Image Interpretation Keys

Identifying cover types that represent basic materials on the ground involves the recognition of patterns and the interpretation of colours or shades. The colour, or spectral response, of natural and man-made features is often similar because of the common characteristics of the materials. Trees, whether planted artificially or occurring naturally, show the same colour. The patterns formed by the plants, however, usually make it possible to distinguish between them due to the species mix and the regular pattern of artificially planted forests. Crops are easily detected from their colour and dense pattern. Pavement and many rock types are often indistinguishable by colour alone because soils, sand, and gravel are often used as construction materials. Looking for patterns is the first step in interpreting an image, and the next step is to determine of what material a surface is likely to be composed.

12.2.1.1 Grey Scale and Colour

The brightness and appearance of a cover type in an image depends on the wavelength band of the image, i.e., how much radiation is reflected in that colour or wavelength, and the relative brightness of other features in the same image (contrast). Boundaries between areas of vegetation and areas with no vegetation show up well in the near-visible infrared (NIR) because of the

Fig. 12.7 Digital airphoto showing an agriculture area in North New South Wales, Australia. The difference in brightness indicates if the field is cropped or left as bare

high reflection of vegetation, the relatively lower reflection of other surfaces, and the lack of haze and atmospheric scattering in the infrared. Moreover, water absorbs most NIR, resulting in sharp water-land boundaries.

The visible wavelength bands are useful in the identification of man-made objects, partially because they tend *not* to consist of growing vegetation. The bright response from concrete, brick, and gravel tends to show up well against the dark background of vegetation (in parks, gardens, fields, and pastures). The visible bands are also helpful to identify agriculture crop fields and to monitor crop growth (Figure 12.7). Vegetation absorbs most red lights because chlorophyll in the leaves uses red light (and blue light and, to a lesser extent, green light) to turn nutrients and water into plant matter, and in the process the plant consumes carbon dioxide and releases oxygen into the atmosphere. The stark contrast between bright vegetation in NIR, and bright concrete and rock in red colour, means that a colour composite that includes red and NIR wavelengths is very useful for identifying features.

The mid-infrared bands of Landsat TM imageries are especially useful in rural environments. Soil differences tend to show up in the mid-infrared (MIR) generally because of differences in water and mineral content. Water absorbs radiation in the MIR area of the spectrum. Certain minerals, specifically clays that contain water in their structure, absorb mid-infrared radiation for this reason. The water content of plants will change the amount of MIR radiation that is absorbed. For example, dry grass will reflect much more MIR than the wetter grass. Other minerals such as carbonates absorb radiation only in the longer MIR. The usefulness of the MIR bands is one reason that TM data, even with lower spatial resolution compared with SPOT HRV, are most commonly used in rural areas.

12.2.1.2 Pattern and Texture

The pattern of naturally occurring features tends to be somewhat irregular and random, while those of man-made features are more geometrical and regular. Figure 12.8 shows examples of typical spatial patterns and textures for natural and artificial features.

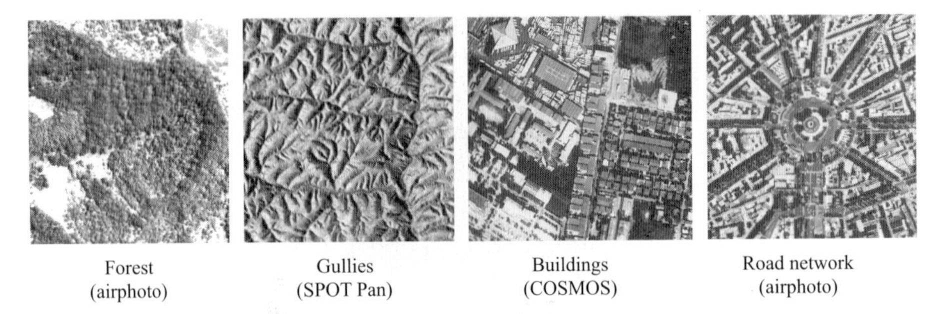

Fig. 12.8 Some examples of spatial patterns and textures for natural and artificial features

12.2.1.3 The Shape of Objects

The shape of an object describes its external form or configuration. Cultural objects tend to have geometrical shapes and distinct boundaries, whereas natural features tend toward irregular shapes with ill-defined boundaries. Figure 12.9 shows a well-known cultural feature that is so unique that it takes little effort to identify. Many natural features also have distinctive shapes, such as sand dunes, volcanic cinder cones, alluvial fans, and riverbeds (Figure 12.10).

12.2.1.4 The Spatial Context

Certain objects are "geographically" linked to other objects, so that identifying one tends to indicate or confirm the other. The spatial context is one of the most helpful clues for identifying cultural features that are composed of aggregate components. For example, a bright area typical of concrete materials next to a runway would most likely indicate an airport

Fig. 12.9 COSMOS imagery showing a pyramid near Cairo, Egypt

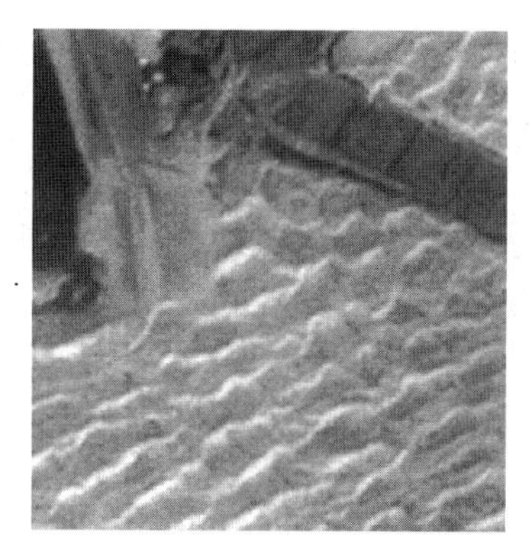

Fig. 12.10 SPOT Panchromatic image of Yulin City, Shaanxi Province, China, showing distinctive shapes of sand dunes

terminal, whereas scattered lighter spots on water can reasonably be supposed to be sea vessels. The spatial context for natural features, however, requires some knowledge about geography. For example, Figure 12.6 shows an ancient lake that is now dry. On its lee side (right side on the image) is a bright strip that indicates a sandbank (also called "lunette") that is the result of a long-history of aeolian sedimentation.

12.3 Image Processing for Thematic Information Extraction

Image processing can be done manually through visual interpretation. This is a rather slow process, and it is now gradually being computerized. The central idea behind digital image processing is quite simple. The digital image is fed into a computer one pixel at a time with its brightness value, otherwise known as its digital number (DN). The computer will perform some calculations for each pixel and store the result. There are many manipulations possible, but we will confine our discussion in this chapter to image arithmetic and image classification.

12.3.1 Image Arithmetic

One important aspect of multispectral images is their ability to detect the differences between spectral bands, which can then be compared qualitatively or quantitatively with the spectral signatures of different materials found on Earth (refer to Figure 12.5). Often the image processing functions using multispectral bands can be presented in a way similar to variables in a mathematical equation. For example, one image band can be added on, or subtracted from, another band on a pixel-by-pixel basis. This approach, therefore is commonly called *image arithmetic*.

12.3.1.1 Band Ratio and the Topographic Effect

Among a large variety of image arithmetic functions, band ratio is arguably the most commonly used. Typical applications include the removal of topographic effects on images and the detection of different cover types, including the derivation of vegetation indices.

The amount of illumination that a point on a surface receives is a function of the angle at which the light is hitting the slope. A typical area on the surface of Earth receives the most light when the sun is directly overhead. The amount of light that is reflected back to a sensor, thus, is a function of not only the properties of the surface material but also the angle of illumination. This property is known as the *topographic effect* and its effect is shown in Figure 12.11.

Two stands of identical material will appear different if they are receiving differing amounts of illumination. Even under different light conditions, if the only variable affecting the response is illumination due to slope angle, then the ratio between matching pixels in each band will remain the same. This is because the percent of the total light hitting the surface remains the same for each band, even if the absolute amount is different (Figure 12.12). In reality, atmospheric effects, variation in the cover type response to the angle of illumination, and sensor calibration make quantitative analysis using band ratios difficult. The identification of spectral features using ratios is more promising.

Although band ratios do not entirely remove the effects of illumination, they do reduce its effect to the point that spectral differences between cover types can be more easily identified. In many cases two surfaces look similar in most bandwidths. A particular characteristic, such as selected absorption due to a mineral component, may impart a subtle difference between two regions in one band. These subtle differences can be enhanced using a ratio.

Band ratios can be used in any instance where the absorption in one band is different from that in another band. Stressed vegetation can be separated

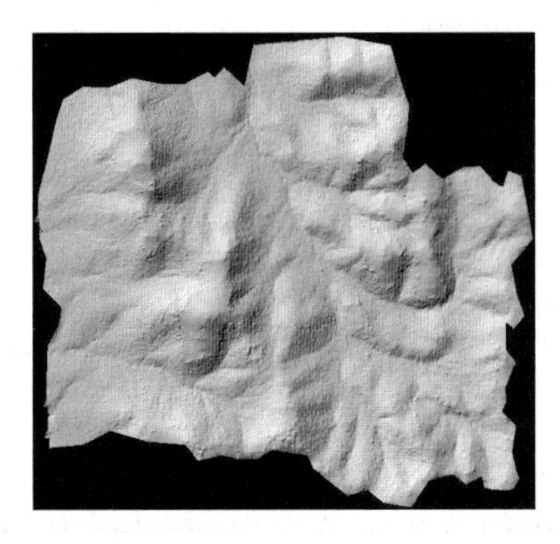

Fig. 12.11 A simulated image of the topographic effect with a constant cover type

DN on slope facing away from sun				DN on slope facing sun			
Unit	Red	NIR	Red/NIR	Unit	Red	NIR	Red/NIR
A	45	60	0.75	A	60	80	0.75
B	20	40	0.50	B	30	60	0.50

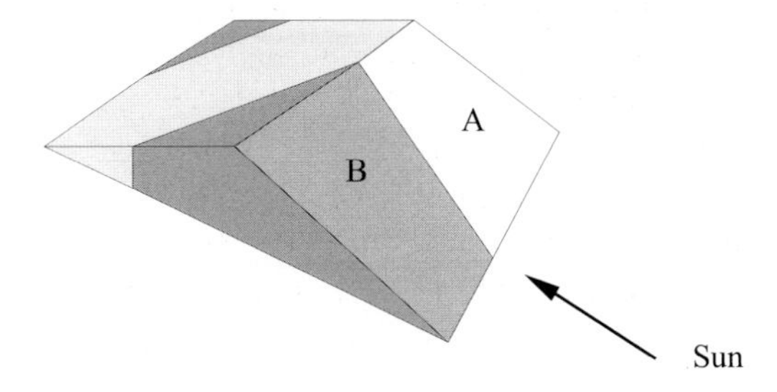

Fig. 12.12 The ratio between areas of the same cover type that are illuminated differently should be constant (sun illumination from the right)

from healthy vegetation in many cases; algae can be found in turbid water; and the parent materials of soils can be traced. One typical example of this is based on the sharp difference in spectral signatures between bare soil and green vegetation. The former presents a near linear spectral curve in the region of visible and near infrared (NIR), while the latter typically has a low reflectance value in red (R) and a very high reflectance value in the NIR area. Therefore the R/NIR ratio should distinguish the bare soil (with a high value) and green vegetation (with a low value). A summary of band ratios that are useful for discriminating various surface materials is as follows:

Soil, Vegetation:	$\frac{R}{NIR}$	Clays:	$\frac{TM5}{TM7}$
FeO, Hydroxides:	$\frac{R \text{ or } G}{B}$ or $\frac{R}{G}$	Plant stress:	$\frac{TM5}{TM7}$ or $\frac{TM3}{TM5}$ or $\frac{MSS3}{MSS4}$

12.3.1.2 Vegetation Indices and their Interpretation

A very popular application of the band ratio function is the vegetation indices. Vegetation can be separated from other cover types because it has characteristic absorption features in visible wavelengths, and high reflectivity in near-visible infrared wavelengths. The spectral curve of vegetation presented in Figure 12.5 is clearly a different shape than that of soil. On the red TM images presented in Figure 12.13, the fields with green crops appear to be slightly darker and are not easily distinguishable from those without crops. On the other side, the high reflectivity of green vegetation is detected by the NIR sensor so that the cropped fields are clearly distinguishable (Figure 12.14).

When vegetation becomes stressed, absorption decreases in visible wavelengths and increases in near-visible infrared wavelengths (NIR). Additionally, the higher the density of broad-leaf plants, the more distinct will be the difference between visible and NIR.

Fig. 12.13 Red TM channel (band 3)

Fig. 12.14 NIR TM channel (band 4)

To date, there are many different kinds of vegetation indices that have been developed and reported in the literature. Most of them are based on the fundamental comparison between the red band and the NIR band of remotely sensed images. The simplest index using these features is an NIR and red band differential (DVI).

$$DVI = NIR - R \tag{12.1}$$

If there is substantial topographic effect due to rugged terrain, the variation in illumination (due to slope orientation) will cause the resulting difference to vary substantially throughout the image.

The topographic effect can be minimised with a ratio, as discussed in section 12.3.1.1. The current most widely used vegetation index is the *Normalised Difference Vegetation Index* (NDVI) which can be expressed as:

$$NDVI = \frac{NIR - R}{NIR + R} \tag{12.2}$$

The NDVI is primarily based on the NIR and red ratio, but it normalises the output values within the range of [–1, 1]. This provides advantages not only for reducing the problems related to illumination differences due to topography, but this also makes the result easier to interpret. Practically, when NDVI$\leqslant$0, one can quite comfortably assume that the pixel is not vegetation. The more active (or 'greener') the plant is, the higher will be the NDVI value returned on the image. While NDVI $\rightarrow$ 1, the pixel will most likely be covered by active (green) plants. A comparison between simple NIR/red ratio and NDVI is shown in Figures 12.15 and 12.16.

Quite often in applications, vegetation indices are derived to separate the vegetation cover from bare soils, rocks, urban areas, etc. A common

Fig. 12.15 Simple ratio of NIR/red (TM4/TM3)

Fig. 12.16 NDVI

technique is to compute the vegetation index and then do a density slice of the resulting image so that the areas with different levels of vegetation coverage may be distinguished. Attempts have also been made with variable degrees of success to quantify the vegetation cover by relating ground measurements to the vegetation indices.

12.3.2 Image Classification

Image classification is the process of creating a meaningful digital thematic map from an image data set. The map classifications are derived either from known cover types (wheat, soil) or from algorithms that search the data for similar pixels. Once data values are known for the distinct cover types in the image, a computer algorithm can be used to divide the image into regions that correspond to each cover type or *class*. The classified image can be converted to a land use map if the use of each area of land is known. The term *land use* refers to the purpose for which people put the land to use (e.g. city, national parks, or roads), whereas *cover type* refers to the material that makes up an area (e.g. concrete, soil, or vegetation).

Image classification can be done using a single image data set, multiple images acquired at different times, or even image data with additional information such as elevation measurements or expert knowledge about the area. Pattern matching can also be used to help improve the classification. The discussion here concentrates on the use of a single image data set to create a classified thematic map where each pixel is classified based on its spectral characteristics. The process that would be used for multiple images is essentially the same with perhaps some extra effort needed to match the images. If soil type or elevation are used the algorithm would need to take into account the fact that thematic soil classes need to be treated differently than measured radiance data.

12.3.2.1 Turning Pixel Values into a Thematic Map

Classification algorithms are grouped into two types of algorithms: supervised and unsupervised. With the *supervised classification* the analyst identifies pixels of known cover types and then a computer algorithm is used to group all the other pixels into one of those groups. With the *unsupervised classification* a computer algorithm is used to identify unique clusters of points in data space, which are then interpreted by the analyst as different cover types. The resulting thematic image shows the area covered by each group or class of pixels. This image is usually called a thematic image, or *classified image*.

Figures 12.17 and 12.18 show the processes that are used to create the classified image. Unique cover types are identified either by the computer (clustering) or by the analyst (from the image). If clustering is used, the

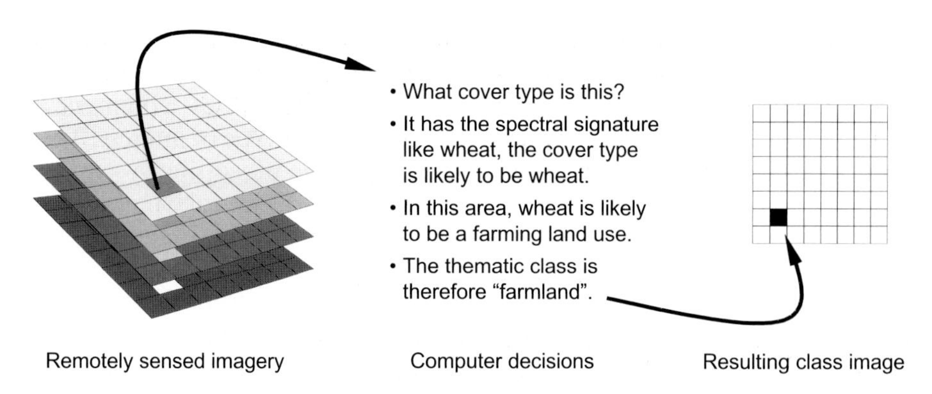

Fig. 12.17 The process of making a classed thematic map from a digital data set

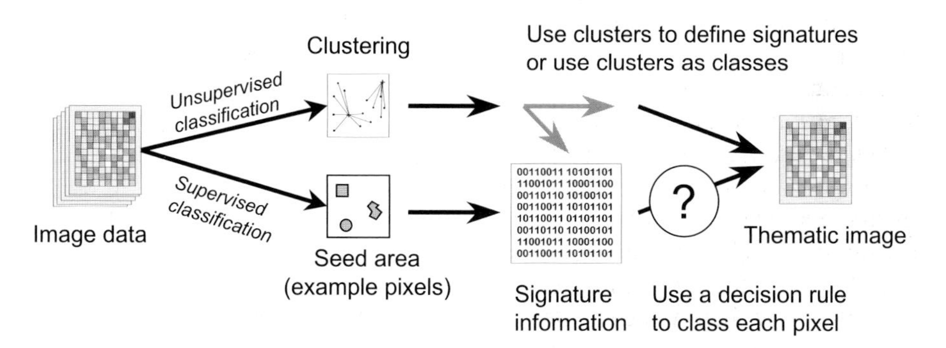

Fig. 12.18 Diagram showing the steps to produce a thematic or classified image

pixels in an image are sorted into groups that are spectrally unique. These can be used either directly as a classified image or to define a set of spectrally unique *signatures* (the statistical description of the class). If the user has chosen example pixels, the pixel samples are used to calculate the signatures of each cover type class (vegetation, sand, etc; signatures will be discussed further). Once signatures have been defined, an algorithm called a decision rule is used to place each pixel from the image data set into one of the classes. The process is often repeated a number of times, adjusting the signatures and decision rule before each run, each time checking the results against areas in the image in which the cover types are known.

12.3.2.2 Supervised Classification

The supervised classification relies on the analyst who provides the 'training' for computers to recognise different cover types. Usually there are three basic steps involved in a typical supervised classification procedure, namely *training*, *classification*, and *output*.

The purpose of the *training* stage is to derive spectral signatures for the cover types of interest to create 'seeds' for classification in the later stage. The analyst identifies representative training areas and develops a numerical description of the spectral attributes of each land cover type of interest. This training can be carried out interactively on the image processing system by selecting 'training areas' in which the pixel DNs of multispectral bands can be statistically analysed to derive a spectral signature of the class (Figure 12.19). Alternatively, one can 'train' the computer by selecting certain DN range in a multi-dimensional spectral space (e.g., on a scattergram) and then examining the corresponding selected areas on the image (Figure 12.20).

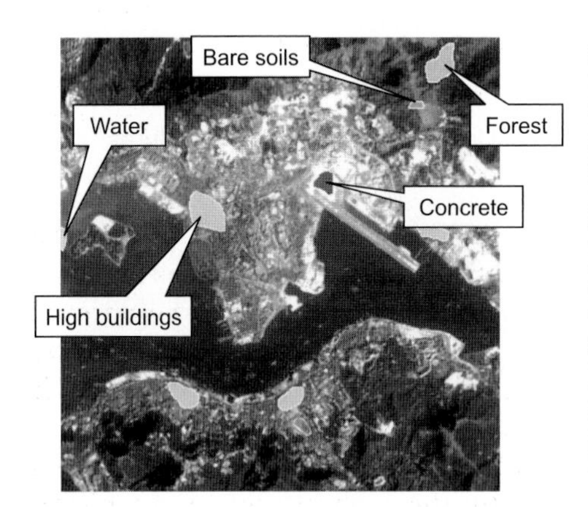

Cover Type	Colour	No. Points
Water	Cyan	3793
Concrete	Purple	975
High buildings	Thistle	1866
Bare soils	Coral	784
Grass slope	Yellow	924
Forest	Green	3122

Fig. 12.19 Training areas are interactively selected on the image for different cover types to derive their spectral signatures for classification

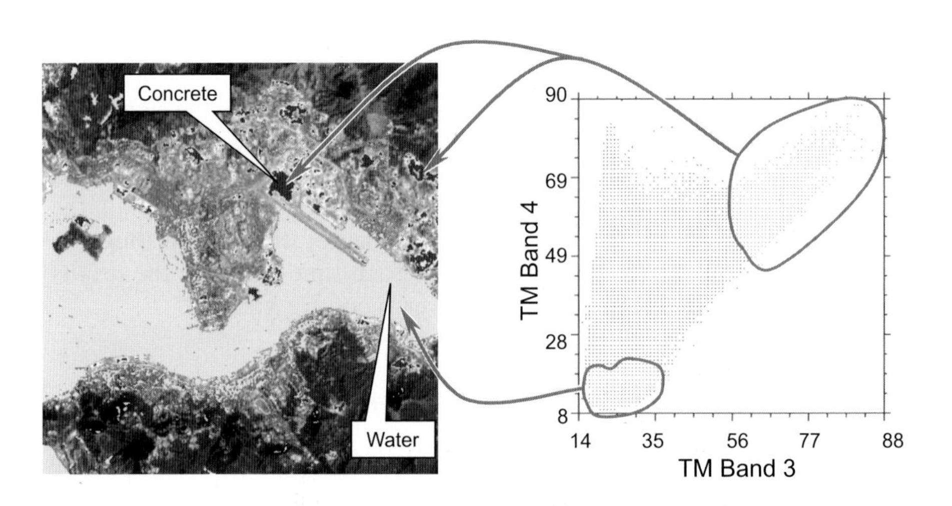

Fig. 12.20 Spectral ranges are selected on the scattergram (right) and the pixels with the selected spectral characteristics are interactively marked as the training areas (left)

In the *classification* stage, each pixel in the image is categorised into the cover class it most closely resembles. If the pixel is not spectrally similar enough to any seed created by the training process, it is then labelled "unknown". The class label (or the theme) assigned to each pixel is then recorded in the corresponding cell of an interpreted data set, or *classified image*.

Today the analyst has a variety of choices in the way to define how 'close' a pixel is to its nearest seed of pre-defined classes. This choice often refers to the selection of *classifiers* that are based on spectral pattern recognition. Numerous mathematical methods of spectral pattern recognition have been developed and extensive discussion of this subject can be found in the literature. For the purpose of this chapter, our discussion touches only the surface of the vast knowledge base about how spectral patterns may be classified into categories, by using limited examples, namely *parallelepiped*, *minimum distance* and *maximum likelihood* classifiers. For ease of presentation, various approaches towards classification are illustrated with a two-band multispectral image. In reality, rarely are just two bands employed in an analysis.

Assume that we sample an image with pixel observations from areas of known cover type (i.e., from the training areas). Each pixel value is plotted on the scattergram that shows various distributions of the spectral response patterns of each cover type to be interpreted in two-dimensional spectral space (Figure 12.21). Our consideration is thus the strategies of using these 'training' spectral response patterns as interpretation keys by which other pixels are categorised into their appropriate classes.

The *parallelepiped classifier* considers the range of values in each class's training set. This range may be defined by the highest and lowest digital number values in each band and appears as rectangular boxes in our two-dimensional scattergram (Figure 12.22). When a pixel lies inside one of the

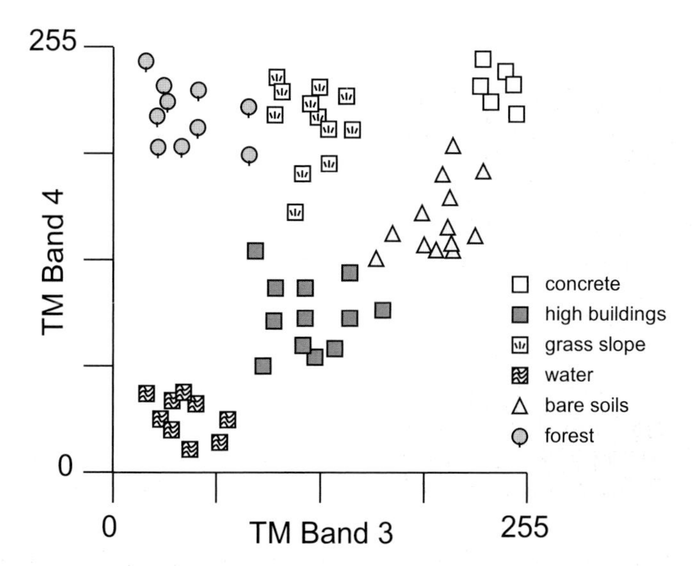

Fig. 12.21 The scattergram illustrating the pixel observations of six cover types

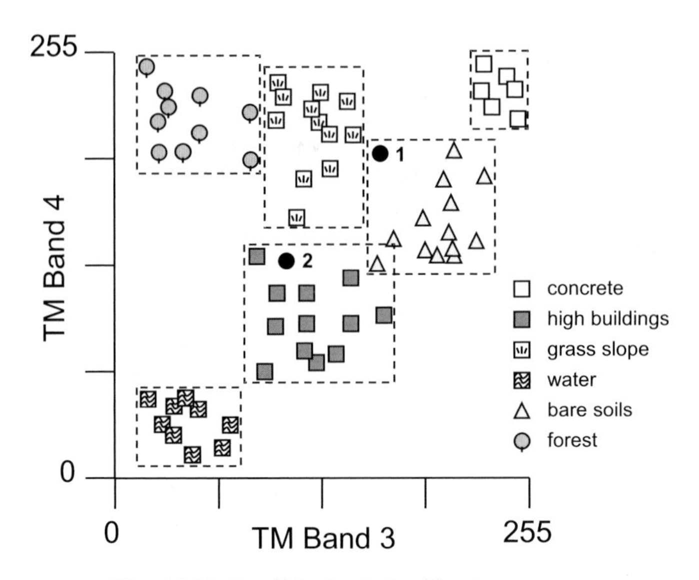

Fig. 12.22 Parallelepiped classification strategy

box, then it is classified into the corresponding class (e.g., point 2 in Figure 12.22). If a pixel lies outside all regions, then it is classified as "unknown".

Difficulties are encountered when class ranges overlap where a pixel has to be classified as "not sure" or it has to be arbitrarily placed in one of the two overlapping classes.

Because spectral response patterns often exhibit correlation, or high covariance, and the rectangular decision regions fit the class training data poorly, the result is confusion for a parallelepiped classifier. For example, point 1 shown in Figure 12.22 would probably be better classified into the class "grass slope" rather than "bare soils" as shown. This problem can be somewhat amended by modifying the single rectangles into a series of rectangles with stepped borders.

The *minimum distance classifier* comprises three steps. First the mean of the spectral value in each band for each class is computed (represented in Figure 12.23 by symbol "+"). Then the distance between the spectral value of an unknown pixel and each of the category means can be computed. Then the pixel is assigned to the "closest class".

The minimum distance classifier is mathematically simple, and it overcomes the poor representation problem of a rectangular decision region used by a parallelepiped classifier. For example, point 1 shown in Figure 12.23 would be correctly classified as "grass slope". This strategy, however, has its limitations. It is insensitive to different degrees of variance in the spectral response data. In Figure 12.23, point 2 would be classified as "concrete" in spite of the fact that the pixel would probably be more appropriate as "bare soils" because of the class's greater variability.

The *maximum likelihood classifier* quantitatively evaluates both the variance and covariance of the category spectral response patterns when

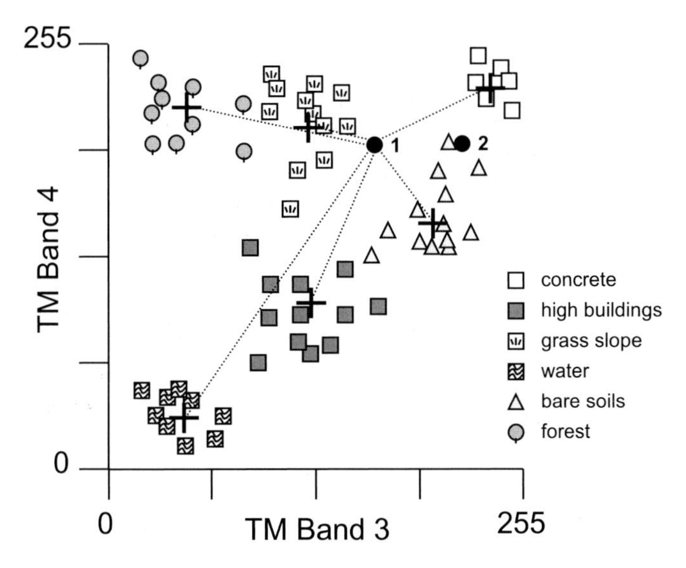

Fig. 12.23 Minimum distance classification strategy

classifying an unknown pixel. To do this, an assumption is made that the pixel spectral cluster forms a normal distribution, which is considered reasonable for common spectral response distributions. Under this assumption, we may compute the statistical probability of a given pixel value being a member of a particular cover class by applying a *probability density function* for each class derived from its training data.

Using the probability density functions, the classifier would calculate the probability of the pixel value occurring in the distribution of the "concrete" class, then the likelihood of its occurring in the "high buildings" class, and so on. After evaluating the probability in each class, the pixel would be assigned to the most likely class that presents the highest probability value, or labelled "unknown" if the probability values are all below a given threshold.

Figure 12.24 shows the probability values plotted on our two-dimensional scattergram where the contour lines are associated with the probability of a pixel value being a member of one of the classes. Basically the maximum likelihood classifier delineates ellipsoidal equal-probability contours, the shape of which shows the sensitivity of the classifier to both variance and covariance. For example, both pixel 1 and 2 would be appropriately assigned to the "grass slope" class and "bare soils" class, respectively.

The principal drawback to maximum likelihood classification is the extensive demand on computation to classify each pixel. When a large number of spectral bands are involved or a large number of classes must be differentiated, the maximum likelihood classifier would perform much slower than the other classifiers described. This drawback was one of the major limitations in the past, but is becoming much less critical today with the rapid development of computer hardware.

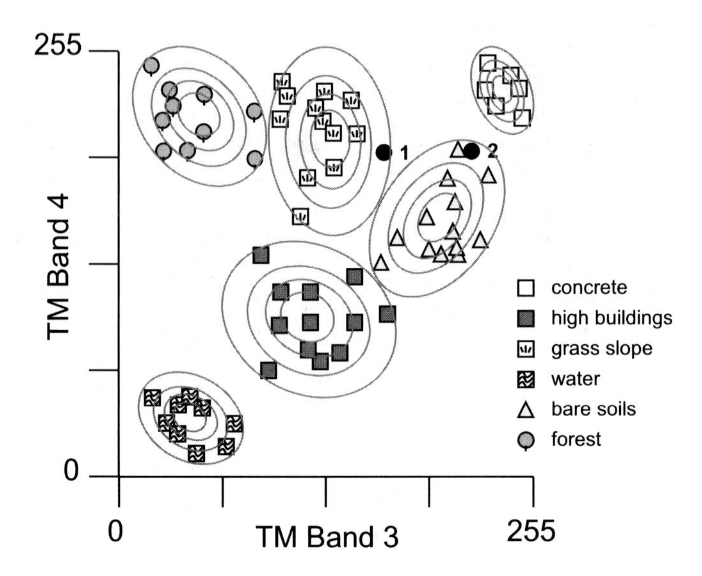

Fig. 12.24 Equal-probability contours defined by a maximum likelihood classifier

In the *output stage*, the results are presented in the form of thematic maps, tables of statistics for the various cover classes, and digital data files suitable for inclusion in a GIS. Since the multispectral classification methods are primarily based on the spectral characteristics with minimum consideration of the spatial extents of the resulting classes, the final classified image is often cluttered with a considerable number of high-frequency spatial variations (Figure 12.25, left image). Because many applications require the classification results to be used as input into a GIS in a form similar to other thematic data layers, often a post-classification process needs to be performed.

The most common demand on a post-classification process is to remove high-frequency spatial variance (or 'noise') from the classified image. This is often achieved by analysing the neighbourhood for each pixel and removing the scattered single pixels ('sieve' process), and then merging the small patches of pixels together to make more continuous and coherent units ('clump' process). The effect of this process is illustrated in Figure 12.25.

12.3.2.3 Unsupervised Classification

Unsupervised classifiers do not utilise training data as the basis for classification. Rather, this kind of classifier involves algorithms that examine the unknown pixels in an image and aggregate them into a number of classes based on the natural groupings or *clusters* present in the image values. The basic assumption here is that values within a given cover type should be close together in the multi-dimensional spectral space, whereas data in different classes should be comparatively well separated.

Fig. 12.25 Classified image before (left) and after (right) post-classification process

Unlike supervised classification, the classes that result from unsupervised classification are *spectral classes*. Because they are solely based on the clusters in the image values, the identity of the spectral classes is not initially known. The analyst must compare the classified data with some form of reference data to determine the identity and informational value of the spectral classes.

Clustering algorithms use predefined parameters to identify cluster locations in data space, and then to determine whether individual pixels are in those clusters. In many algorithms the number of clusters may be defined at the start, while others just use cluster size and separation parameters to control the number of clusters that are found. Figure 12.26 illustrates the type of parameters that can be used to define clusters, and whether pixels belong in that cluster. Clustering algorithms either pass once through the data, grouping pixels during that pass, or they pass through a number of times to adjust and improve the clustering assignments. It is impossible to discuss all forms of clustering in this book; however, most clustering algorithms used in remote sensing software operate in a similar manner.

A typical multiple-pass, or iterative, clustering algorithm works as shown in Figure 12.27. Pass One: (A) Cluster centres are arbitrarily assigned. (B) Each pixel is assigned to the cluster centre nearest to them in data space (spectral distance). (C) The cluster means are then calculated from the average of the cluster members (the middle cluster is shown with grey points) and the pixels are reassigned to the new cluster centres. Pass Two: (D) the process is repeated. The iteration stops when the cluster centres (or means) move by less than a pre-set amount during each iteration. With a number of iterations the location of clusters tend to stabilise as the location of cluster centres between each pass changes less and less.

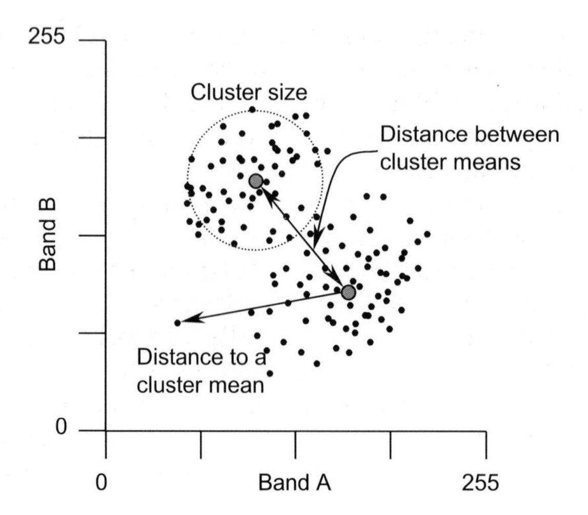

Fig. 12.26 Measures that define a cluster include the size of a cluster and the distance between clusters

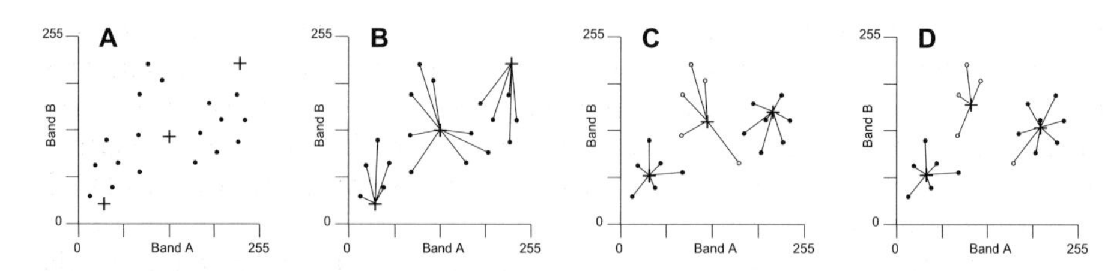

Fig. 12.27 Iterative clustering of points in data space

Algorithms that pass through the image only once tend to be more affected by the initial conditions than iterative algorithms that repeatedly adjust the cluster means. After each pass through the data, cluster means can be calculated along with other measures, such as standard deviation. In addition to simple straight-line distances, statistical measures of distance can be used where the distance to clusters is weighted by the size and importance of that cluster.

The result from unsupervised classification may also need a post-classification process as described above. In addition, because the real-world nature of spectral classes derived from the classification is largely unknown, considerable analysis and interpretation will be required. Often the resulting classes need to be merged into fewer classes to make the classified image more acceptable as a thematic map.

12.3.3 Accuracy Assessment

Classification accuracy analysis is one of the most active research fields in remote sensing. Meaningless and inconclusive assessment of the image classification results sometimes precludes the application of automated land cover classification techniques even when their cost is more favourable with more traditional means of data collection. It is always claimed in the remote sensing community that "a classification is not complete until its accuracy is assessed" [Lillesand and Kiefer, 1994].

One of the most common methods of expressing classification accuracy is the preparation of a classification *error matrix* (or *confusion table*). The error matrix compares the relationship between the known reference data and the corresponding results of the classification. Table 12.4 shows an error matrix. The numbers listed in the table represent the proportion of training pixels, for each cover type, that were correctly and incorrectly labelled by the classifier. It is common to average the correct classifications and regard this as the overall classification accuracy (in this case 81%), although a better measure globally would be to weight the average according to the areas of classes in the map.

A distinction is made between *omission errors* and *commission errors*. Omission errors correspond to those pixels belonging to the class of interest that the classifier has failed to recognise, whereas commission errors are those that correspond to pixels from other classes that the classifier has labelled as belonging to the class of interest. The former refers to columns of the error matrix, whereas the latter refer to rows. For example, in the case presented in Table 12.4, the omission error for the "concrete" class is $(0+3+3+0+0)/71 = 8\%$, whereas the commission error for the class is $(0+4+6+0+0)/75 = 13\%$.

The *producer's accuracy* shown in Table 12.4 is interpreted as the probability that the classifier has classified the image pixel as, for example, "water" given that the actual class is "water" as indicated by the training data. As a user of the classified image we are more interested in the probability that the actual class is "water" given that the pixel has been labelled as "water" by the classifier (*user's accuracy*). In our case, the producer's accuracy for the "forest" class is $512/581 = 88\%$, whereas the user's accuracy is $512/605 = 85\%$.

For the global assessment of classification accuracy, a measurement called *Cohen's kappa coefficient* (κ) is often employed. The kappa coefficient is a measure that considers significantly unequal sample sizes and likely probabilities of expected values for each class.

Let $x_{i+} = \sum_{j=1}^{n} x_{j,i}$ (i.e. the sum over all columns for row i), and $x_{+j} = \sum_{i=1}^{n} x_{i,j}$ (i.e. the sum over all rows for column j), then

$$\kappa = \frac{d-q}{N-q} \qquad (12.3)$$

where N = total number of samples, d = total number of cases in diagonal cells of the error matrix, and

$$q = \frac{\sum_{k=1}^{n} x_{i+} \cdot x_{+j}}{N}. \qquad (12.4)$$

Table 12.4 An error matrix expressing classification accuracy

Classification results	Training data (known cover types)						
	Water	Concrete	High buildings	Bare soils	Grass slopes	Forest	Row total
Water	**93**	0	2	1	0	0	96
Concrete	0	**65**	4	6	0	0	75
High buildings	2	3	**124**	5	9	12	155
Bare soils	2	3	21	**165**	24	12	227
Grass slopes	0	0	6	16	**201**	45	268
Forest	0	0	8	9	76	**512**	605
Column total	97	71	165	202	310	581	1426

Producer's accuracy

W = 93/97 = 96%
C = 65/71 = 92%
H = 124/165 = 75%
B = 165/202 = 82%
G = 201/310 = 65%
F = 512/581 = 88%

User's accuracy

W = 93/96 = 97%
C = 65/75 = 87%
H = 124/155 = 80%
B = 165/227 = 73%
G = 201/268 = 75%
F = 512/605 = 85%

Overall accuracy = (93 + 65 + 124 + 165 + 201 + 512)/142 = 81%

$\kappa = (1160 - 365.11)/(1426 - 365.11) = 0.749$

The optimal κ score is 1.0 (perfect classification). In our case, $N = 1426$, $d = 1160$, $q = 365.11$, and $\kappa = 0.749$.

12.4 Summary

Digital remote sensing imagery is widely recognised as one important source for spatial information technology. To maximise its potential benefit, however, the images need to be correctly interpreted, classified, and integrated with a GIS operating environment, so that the support for real-time decision making can be delivered.

This chapter discussed the considerations and techniques of information extraction from digital images. The discussion largely focused on the remotely sensed images from various satellite platforms. The majority of such images have a common nature of multispectral capabilities, but with variable spatial resolutions. The spectral, spatial, and temporal characteristics of remotely sensed images and some typical natural and artificial features were reviewed in order to provide a general background for the described techniques.

The discussion on methodology focused on two major areas, namely, feature identification and image interpretation, and image processing for thematic information extraction. The former described keys and methods to be employed in recognising natural and cultural features on Earth's surface, which may be made by different materials. The latter discussed the computer-based image processing techniques for extracting thematic information from digital images.

It is important to understand that this chapter has only 'scratched the surface' of the vast knowledge base of interpretation and machine processing of digital images. Within the scope of this volume, it has been impossible to cover the complete topic. Interested readers may find themselves getting lost in the large literature base of remote sensing technology, but the few references listed below may provide quite useful initial help.

Current Trends in Geographical Data Acquisition: An Epilogue

Zhilin Li and Yong-Qi Chen

The importance of geographical data acquisition has been widely recognized, and it is often reported that about 80% of the budget of a GIS project is spent on data. If there is no data for a project area, then one needs to acquire new sets of data, leading to a process called new survey. If some data exists but some of them are out of date, then there is a need of data revision, leading to a process called re-survey. After the base data sets are available for most of the area, data revision will become an issue.

Looking back at the development of data acquisition methodology, the following trends are recognized:

1) From single sensor to multi-sensor integration;
2) From static mode to dynamic mode;
3) From multi-stage process to real-time;
4) From manual operation to full automation; and
5) From measurement system to monitoring system.

From Single-Sensor to Multi-Sensor Integration

At the early stage of development, diversity is a word which should be used for the description of geographical data acquisition methodology. Aerial photogrammetry replaced ground-based surveying for large area, especially medium and small scale, mapping. Since the 1970s, spaceborne imaging sensors are widely used for the acquisition of geometric and thematic information.

Global positioning system (GPS) is another new development. In recent years, GPS receivers have been placed onboard so that the position and orientation parameters of the aerial camera can be recorded while the photography is being taken. This integration leads to a new jargon called "GPS photogrammetry". Similar integration has also happened in close-range photogrammetry. In this case, GPS receivers and CCD cameras are mounted on a van. The CCD cameras are used to take stereo images and GPS receivers are used to determine the position and orientation of the

CCD camera. This integration leads to another new jargon called "mobile mapping".

From Static Mode to Dynamic Mode

Ground-based surveying employs theodlite, total station, level and other instruments. With them, all the measurements are carried out statically. Currently, GPS receivers can operate either in a static mode or dynamic mode (or kinematic). In dynamic mode, GPS receivers may be carried by surveyors or mounted on a vehicle, as in the case of mobile mapping and GPS photogrammetry.

In fact, most of image acquisition systems operate in dynamic model. Aerial photography and satellite images are taken while the platform, and thus the sensors, are moving at high speed. Even ground-based imaging systems have been mounted on land vehicles to take photographs or video images in a dynamic mode, as in the case of mobile mapping systems.

From Manual Operation to Full Automation

When a ground-based survey is carried out, a set of data is traditionally recorded on paper. This set of data is then keyed into the computer for mapping and other applications. With current developments, survey data is recorded electronically into a data-logger. This data logger can then be connected to a computer and data can be read by application programs. This is towards full automation.

In photogrammetry, many of the processes have been automated, from reconstruction of stereo models (through relative orientation and absolute orientation) to measurement. In measurement, image-matching techniques have been developed to derive the three-dimensional coordinates of feature points and feature extraction algorithms have been developed to derive the two-dimensional coordinates of well-defined linear features. These functions have already been implemented in digital photogrammetric workstations (DPW). In the acquisition of thematic information from digital images, classification algorithms have been developed so that digital images can be automatically interpreted. These functions have been implemented in remote sensing image-processing systems.

From Multi-Stage Process to Real-Time

A typical multi-stage process is aerial photogrammetry. Photographs are taken by a camera mounted onboard an airplane. Then these photographic films are processed and diapositives (the positive images on transparent films) made. The diapositives are then used in photogrammetric plotters for reconstruction of stereo-models. After that, measurement can be made on these stereo-models and the output of measurement could be in graphic form and/or in digital form at later stage.

With current developments, CCD cameras are used to replace analogue camera so that digital images are taken. These digital images can, through telecommunication technology, be transmitted to a computer in the office in real-time or nearly real-time. Onboard the airplane, GPS receivers are also mounted so that the position and orientation parameters of the CCD camera are also recorded. With these parameters known, the digital images can then be used for different applications. Alternatively, digital computers can also be mounted on airplanes so that digital images can be directly transmitted to application systems. This development leads to another new jargon called "Airborne GIS".

From Acquisition Systems to Monitoring Systems

At an early stage of development, data acquisition is carried out by systems designed for just performing measurements such as total stations. However, spaceborne systems offer not only measurement but also repeated coverage of the same area at a regular time interval. The spatial resolution of images has been improved dramatically, from MSS on Lansat-1 (79 m) to TM (35 m), SPOT (20 m for MSS and 10 m for panachromatic) to IKONOS (1 m). In the near future, sub-metre resolution space images will be available. The temporal resolution of space images is also reduced dramatically from 16 days to 2 days.

The spaceborne radar system is another interesting development for the acquisition of geographical data for monitoring. The system operates on microwaves so that it can penetrate cloud. In this way, it can operate under all weather conditions. Synthetic aperture radar (SAR) interferometry, called InSAR, is a more recent development of space remote sensing technology. It employs two SAR images to form interferograms. InSAR can be used to measure the position of features as well as the deformation of objects.

References

Abdel-Aziz, Y. L. and H. M. Karara (1971) Direct linear transformation from comparator coordinate into object space coordinates in close range photogrammetry. Proceedings of ASP Symposium on Close-range photogrammetry

Abidin, H. Z. (1994) On-the-Fly Ambiguity Resolution. GPS World, April, pp. 40–50

Ackermann, F., C. Braun, W. Förstner, C. Fuchs, E. Gülch, R. Laing, F. Lang, M. Knabebschuh, P. Krzystek, T. Löcherbach, T. Schenk, W. Schickler, U. Weidner (1995) Second course in digital photogrammetry. Landesvermessungsamt Nordrhein-Westfalen, Bonn

Ackermann, F. (1999) Airborne laser scanning-present status and future expectations. ISPRS Journal of Photogrammetry and Remote Sensing, Vol. 54, Issue 2–3, pp. 64–67

Agfa-Gevaert (1996) A guide to digital photography. Theory and basics. Agfa-Gevaert, Belgium

Allan, A. L. (1993) Practical Surveying and Computations, 2nd edn. Butterworth-Heinemann Ltd.

American Society of Civil Engineers (1996) Photogrammetric Mapping. ACSE Press, New York

ASPRS (1997) Manual of Remote Sensing, 3rd edn., American Society of Photogrammetry and Remote Sensing. Bethesda, Maryland

Ashkenazi V., C. H. J. Chao, W. Chen, C. J. Hill, and T. Moore (1997) A New High Precision Wide Area DGPS System. The Journal of Navigation, Vol. 50, No. 1, January 1997, pp. 109–119

Atkinson, K. B. (ed.) (1996) Close range photogrammetry and machine vision, Whittles Publishing, Caithness

Avery, T. E. and G. L. Berlin (1992) Fundamentals of Remote Sensing and Airphoto Interpretation, 5th edn, Prentice-Hall, Upper Saddle River, NJ

Baltsavias, E. P. (1996) Digital ortho-images – a powerful tool for the extraction of spatial- and geo-information. ISPRS Journal of Photogrammetry and Remote Sensing, Vol. 51, Issue 2, pp. 63–77

Baltsavias, E. P. (1999) A comparison between photogrammetry and laser scanning. ISPRS Journal of Photogrammetry and Remote Sensing, Vol. 54, Issue 2–3, pp. 83–94

Blachut T. J., A. Chrzanowski and J. H. Saastamoinen (1979) Urban Surveying and Mapping. Springer-Verlag

Bomford G. (1980) Geodesy, 4th edn. Clarendon Press, Oxford

Brinker R. and R. Minnick (eds.) (1995) The Surveying Handbook. Van Nostrand Reinhold, New York

Chen, Y. Q. and Y. C. Liu (1997) Correction for the Seabed Distortion caused by the Angular Beamwidth of Echo Sounders. The Hydrographic Journal, Hydrographic Society, UK, No. 84, pp. 15–19

Chen Y. Q. (1997) Development in Engineering Surveying. The Geomatics Journal of Hong Kong, Vol. 1, No. 1, pp. 1–8

Ciciarelli, J. A. (1991) A Practical Guide to Aerial Photography. Van Nostrand Reinhold, New York

Chapman, D. and A. Deacon (1998) Panoramic imaging and virtual reality–filling the gaps between the lines. ISPRS Journal of Photogrammetry and Remote Sensing, Vol. 53, Issue 6, pp. 311–319

Clarke, T. A. (1995) An analysis of the prospects for digital close-range photogrammetry. ISPRS Journal of Photogrammetry and Remote Sensing, Vol. 50, Issue 30, pp. 4–7

Cracknell, A. P. and L. W. B. Hayes (1991) Introduction to Remote Sensing. Taylor and Francis, London

Cross, P. A., J. R. Hollwey and L. G. Small (1984) Geodetic Appreciation. North East London Polytechnic, Department of Land Surveying

De Loach, S., D. Wells and D. Dodd (1995) Why On-the-Fly? GPS World, May 1995, pp. 53–58

Derenyi, E., Y. C. Lee, G. W. Moore and A. Chow (1993) The speed and accuracy of semi-automated digitizing. Proceedings of the 7th Annual Symposium on Geographic Information Systems in Forestry, Environment and Natural Resources Management, Vol. 2, February 15–18, Vancouver, B. C., pp. 1011–1019

Ebner, H., B. Hofmann-Wellenhof, P. Reiss and F. Steidler (1980) HIFI–A minicomputer program package for height interpolation by finite elements. International Archives for Photogrammetry and Remote Sensing, Vol. 23, Part IV, pp. 202–241

Elmasri, R. and S. B. Navathe (1994) Fundamentals of Database Systems. Benjamin/Cummings, California

Ferguson, J. (1990) How the flat earth got round: an ancient history. Ontario Land Surveyor, Summer, pp. 8–9

Ferguson, J. (1991) What is height anyway? Ontario Land Surveyor, Spring, pp. 17–19

Fowler, R. (1992) All about coordinate systems. Northpoint, Summer, pp. 20–24; Fall, pp. 24–28

Garland, G. D. (1977) The earth's shape and gravity. Pergamon Press, London

Ghosh, S. K. (1988) Analytical photogrammetry. Pergamon Press, New York

Ghosh, S. K. (1968) Theory of stereophotogrammetry. Dept. of Geodetic Science, Ohio State University, Columbus

Gonzalez, R. C. and R. E. Woods (1992) Digital Image Processing. Addison-Wesley Publishing Company

Graham, R. W. and R. E. Reid (1986) Manual of aerial photography. Focal Press, London

Greve, C. (1996) Digital photogrammetry: an addendum to the Manual of photogrammetry. American Society of Photogrammetry and Remote Sensing, Falls Church

Gurney, C. M. (1981) The use of contextual information to improve land cover classification of digital remotely sensed data. Int. J. Remote Sensing, Vol. 2, No. 4, pp. 379–388

Haala, N. and C. Brenner (1999) Extraction of buildings and trees in urban environments. ISPRS Journal Of Photogrammetry And Remote Sensing, Vol. 54, Issue 2–3, pp. 130–137

Hallert, B. (1960) Photogrammetry, basic principles and general survey. McGraw-Hill, New York

Han S. and C. Rizos (1997) GPS Ambiguity Resolution Techniques. Reliable Solutions still a Challenge. GIM Journal, November, pp. 31–33

Harris, P. M. and S. J. Ventura (1995) The integration of geographic data with remotely sensed imagery to improve classification in an urban area. Photogrammetric Engineering and Remote Sensing, Vol. 61, No. 8, pp. 993–998

Harvey, B. R. (1990) Practical Least Squares and Statistics for Surveyors, Monograph 13, School of Surveying. University of New South Wales, Sydney, Australia

Heimberg F. and G. Seeber (1994) Some Considerations and Developments to the Use of Differential GPS in Marine Geodesy. Marine Geodesy, Volume 17, pp. 121–138

Higgins, M. (1987) Transformation from WGS84 to AGD84, an Interim Solution. Department of Geographic Information, Old, Australia

Hoffmann-Wellenhof B., B. Lichtenegger and J. Collins (1992) GPS, Theory and Practice. Springer-Verlag, Wien
Hurn, J. (1989) GPS – A Guide to Nexr Utility, Trimble Navigation Ltd.
Hydrographer of the Navy (1965–1970) Admiralty Manual of Hydrographic Surveying. London, UK
Ingham, A. E. (1974) Sea Surveying. Wiley, London, New York
Ingham, A. E. and V. J. Abbott (1992) Hydrography for the Surveyor and Engineer. 3rd edn. Blackwell Science Ltd., Oxford, UK
International Hydrographic Organization (1997). IHO Standards for Hydrographic Surveys. Special Publication No. 44, 4th edn., Monaco
Jordan, Eggert and Kneissl (1996) Die Entfernungsmessung mit elektromagnetischen Wellen und ihre geodätischen Anwendungen. Handbuch der Vermessungskunde, 10. Auflage, Bd. IV, J.B. Metzlersche Verlagsbuchhandlung, Stuttgart 1966 (German)
Junkins, D. and G. Garrard (1998) Demystifying reference systems: A chronicle of spatial reference systems in Canada. Geomatica, Vol. 52, No. 4, pp. 468–473
Kalafus, R. M., A. J. van Dierendonck and N. A. Pealer (1986) Special Committee 104 Recommendations For Differential GPS Service. Global Positioning System, Vol. III. The Institute of Navigation. Washington D. C.
Karara, H. M. (1989) Manual of non-topographic photogrammetry, American Society of Photogrammetry, Falls Church
Langley, R. B. (1990) Why is the GPS signal so complex? GPS World, May/June, pp. 56–59
Langley, R. B. (1992) Basic geodesy for GPS. GPS World, Feb., pp. 44–49
Leick, A. (1993) Geodesy in review. ACSM Bulletin, Jan./Feb., pp. 45–48
Leick, A. (1995) GPS Satellite Surveying. John Wiley & Sons, New York
Lillesand, T. M. and R. W. Kiefer (1994) Remote Sensing and Image Interpretation, 3rd edn., John Wiley & Sons, New York
Macmillan, D. H. (1966) Tides. CR Books, London, UK
Maling, D. H. (1991) Coordinate systems and map projections for GIS. In: D. J. Maguire, M. F. Goodchild, and D. Rhind (eds.) Geographical Information Systems, Principles and Applications, Volume 1, pp. 135–146. Longman Scientific and Technical
Maling, D. H. (1968) The terminology of map projections. International Yearbook of Cartography, Vol. 8, pp. 11–65
Maling, D. H. (1973) Coordinate Systems and Map Projections. George Philip, London
Merritt, L. (1997) 3D modeling from photos: build 3D models fast using desktop photogrammetry. Elizabethton, Tennessee
Mikhail, E. M. (1976) Observations and Least Squares. IEP, New York
Moffitt, F. H. and E. M. Mikhail (1980) Photogrammetry. Harper & Row, New York
Mok, E. (1995) Applying GPS positioning technique in GIS data collection: GPS positioning – How does it work (I)? GIS Asia/Pacific, Vol. 1, No. 2, pp. 36–38
Mok, E. (1995b) Applying GPS positioning technique in GIS data collection: GPS positioning – How does it work (II)? GIS Asia/Pacific, Vol. 1, No. 3, pp. 28–31
Mok, E. (1995c) Understanding the error sources in GPS positioning. GIS Asia/Pacific, Vol. 1, No. 3, pp. 28–32
Mok, E. (1996) Practical Considerations for GPS positioning. GIS Asia Pacific, Vol. 2, No. 1
Olsen, N. T. (1993) Understanding the differences between an ellipsoid, a geoid, and a spheroid. Geodetic Info Magazine. May, pp. 64–65
Philipson, W. R. (ed.) (1997) Manual of Photographic Interpretation, 2nd edn. American Society for Photogrammetry and Remote Sensing, Bethesda, Maryland
Pugh, D. T. (1987) Tides, surges and mean sea level. Wiley, Chichester, New York
Retscher, G. (1997) Application of GPS in Marine Geodesy. Papers presented to the conference PACON 97 Symposium on Resource Development – Environment Issues and the Sustainable Development of Coastal Waters at Hong Kong, August 6–8, 1997
Retscher, G. (1997b) Powerlines and Transmitting Stations Impair RTK-GPS. GIM Journal, November, pp. 65–67

Retscher, G. and F. N. Teferle (1998) Development of a low-cost hydrographic surveying system for inshore and coastal waters. In: Papers presented at the Canadian Hydrographic Conference, March 10–12, 1998, Victoria, British Colombia, Canada
Retscher, G. (1999) RTK-GPS Positioning and Navigation in Marine Geodesy. The Geomatics Journal of Hong Kong, Vol. 1, No. 2, pp. 39–48
Richards, J. A. (1993) Remote Sensing Digital Image Analysis: An Introduction, 2nd edn. Springer-Verlag, Berlin
Schofield, W. (1993) Engineering Surveying, 4th edn. Butterworth-Heinemann Ltd.
Schowengerdt, R. A. (1997) Remote Sensing, Models and Methods for Image Processing, 2nd edn. Academic Press, London
Seeber, G. (1993) Satellite Geodesy. Walter de Gruyter, Berlin, New York
Smith, J. R. (1997) Introduction to Geodesy. Wiley, Inc.
Slama, C. C. (ed.) (1980) Manual of photogrammetry. American Society of Photogrammetry, Falls Church
Snyder, J. P. and P. M. Voxland (1989) An Album of Map Projections. U.S. Geological Professional Paper 1453. USGS
Spradley, L. H. (1996) Costs of softcopy orthophoto bases for GIS projects. ISPRS Journal of Photogrammetry and Remote Sensing, Vol. 51, Issue 4, pp. 182–187
Stolz, A. (1994) An Introduction to Geodesy. Monograph 16, School of Geomatic Engineering, The University of New South Wales
Steinvall, O., J. Banic and M. Alfredsson (1997) Airborne Laser Hydrography: A vital part of future coastal mapping. Hydro International, GITC, Netherlands, Vol. 1, No. 3, pp. 25–27
Survey & Mapping Office, Lands Department (1995) Explanatory Notes on Geodetic Datums in Hong Kong
Thom, C. and I. Jurvillier (1993) Experiences with a digital aerial camera at Institut geographique National (France). In: Fritsch, D. and Bobbie, D. (eds.) Photogrammetric Week'93. Wichmann. pp. 73–79
Thomson, S. (1996) The Future of Commercial DGPS. The Hydrographic Journal, No. 82, October 1996, pp. 3–8
United Nations (1979) Report of the group of experts on hydrographic surveying and nautical charting. 2nd United Nations Regional Cartographic Conference for the Americans, Mexico City
Van Diggelen F. and W. Martin (1997) GPS + GLONASS RTK. A Quantum Leap in Performance. GIM Journal, November, pp. 69–71
Vincenty, T. (1989) The flat earth concept in local surveys. Surveying and Mapping, Vol. 49, pp. 101–2
Warner, W. S., R. W. Graham and R. E. Read (1996) Small format aerial photography, Whittles, Scotland
Weidner, U. and W. Forstner (1995) Towards automatic building extraction from high resolution digital elevation models. ISPRS Journal of Photogrammetry and Remote Sensing, Vol. 50, Issue 4, pp. 38–49
Wells, D. (ed.) (1987) Guide to GPS Positioning. Canadian GPS Associates, Fredericton, Canada
Wells, D. (1996) Overview of Hydrography. Lecture Notes. Department of Geomatics Engineering, University of New Brunswick, Fredericton, Canada
Wells, D. (1996b) Underwater Acoustics. Lecture Notes. Department of Geomatics Engineering, University of New Brunswick, Fredericton, Canada
Wells, D. (1997) Ocean Mapping Status and Trends. The Geomatics Journal of Hong Kong. Vol. 1, No. 1, pp. 19–32
Wilhelm, H., W. Zurn, and H.-G. Wenzel (1997) Tidal Phenomena. Springer, Berlin, New York
Wolf, P. R. and Ghilani, C. D. (1997) Adjustment Computations: Statistics and Least Squares in Surveying and GIS. Wiley, New York
Wolf, P. R. (1983) Elements of photogrammetry, 2nd edn. McGraw-Hill, Singapore

Wray, T. (1974) The seven aspects of a general map projection. Cartographica Monograph, Vol. 11, 72 pp.

Zhang, W., J. Albertz, and Z. Li (1994) Digital orthoimage from air-borne scanner imagery. International Archives for Photogrammetry and Remote Sensing, Vol. 30, Part 3/2, pp. 945–950

Zhang, Y. (1999) Optimisation of building detection in satellite images by combining multispectral classification and texture filtering. ISPRS Journal of Photogrammetry And Remote Sensing, Vol. 54, Issue 1, pp. 50–60

Subject Index

SpringerGeosciences

Bernhard Hofmann-Wellenhof, Herbert Lichtenegger, James Collins

Global Positioning System

Theory and Practice

Fifth, revised edition
2001. Approx. 390 pages. Approx. 40 figures.
Softcover DM 86,–, öS 598,–
(recommended retail price)
ISBN 3-211-83534-2

This new edition accommodates the most recent advances in GPS technology. Updated or new information has been included although the overall structure essentially conforms to the former editions.
The textbook explains in comprehensive manner the concepts of GPS as well as the latest applications in surveying and navigation. Description of project planning, observation, and data processing is provided for novice GPS users. Special emphasis is put on the modernization of GPS covering the new signal structure and improvements in the space and the control segment.
Furthermore, the augmentation of GPS by satellite-based and ground-based systems leading to future Global Navigation Satellite Systems (GNSS) is discussed.

A-1201 Wien, Sachsenplatz 4–6, P.O.Box 89, Fax +43.1.330 24 26, e-mail: books@springer.at, Internet: **www.springer.at**
D-69126 Heidelberg, Haberstraße 7, Fax +49.6221.345-229, e-mail: orders@springer.de
USA, Secaucus, NJ 07096-2485, P.O. Box 2485, Fax +1.201.348-4505, e-mail: orders@springer-ny.com
Eastern Book Service, Japan, Tokyo 113, 3–13, Hongo 3-chome, Bunkyo-ku, Fax +81.3.38 18 08 64, e-mail: orders@svt-ebs.co.jp